MW00607156

METHODS in BRAIN CONNECTIVITY INFERENCE THROUGH MULTIVARIATE TIME SERIES ANALYSIS

FRONTIERS IN NEUROENGINEERING

Series Editor
Sidney A. Simon, Ph.D.

Published Titles

Methods in Brain Connectivity Inference through Multivariate Time Series Analysis
Koichi Sameshima, University of São Paulo, São Paulo, Brazil
Luiz Antonio Baccala, University São Paulo, São Paulo, Brazil

Neuromorphic Olfaction
Krishna C. Persaud, The University of Manchester, Manchester, UK
Santiago Marco, University of Barcelona, Barcelona, Spain
Agustín Gutiérrez-Gálvez, University of Barcelona, Barcelona, Spain

Indwelling Neural Implants: Strategies for Contending with the *In Vivo* Environment
William M. Reichert, Ph.D., Duke University, Durham, North Carolina

Electrochemical Methods for Neuroscience
Adrian C. Michael, University of Pittsburg, Pennsylvania
Laura Borland, Booz Allen Hamilton, Inc., Joppa, Maryland

Forthcoming Titles

Humanoid Robotics and Neuroscience: Science, Engineering, and Society
Gordon Cheng, ATR Computational Neuroscience Labs, Kyoto, Japan

Molecular, Neuropsychological, and Rehabilitation Aspects in Brain Injury Models
Firas H. Kobeissy, University of Florida, Gainesville, Florida

METHODS in BRAIN CONNECTIVITY INFERENCE THROUGH MULTIVARIATE TIME SERIES ANALYSIS

Edited by

Koichi Sameshima

Department of Radiology and Oncology
University of São Paulo
São Paulo, Brazil

Luiz Antonio Baccalá

Department of Telecommunications and Control Engineering
University of São Paulo
São Paulo, Brazil

CRC Press
Taylor & Francis Group
Boca Raton London New York

CRC Press is an imprint of the
Taylor & Francis Group, an **informa** business

CRC Press
Taylor & Francis Group
6000 Broken Sound Parkway NW, Suite 300
Boca Raton, FL 33487-2742

© 2014 by Taylor & Francis Group, LLC
CRC Press is an imprint of Taylor & Francis Group, an Informa business

No claim to original U.S. Government works

Printed on acid-free paper
Version Date: 20140108

International Standard Book Number-13: 978-1-4398-4572-1 (Hardback)

Visit the Taylor & Francis Web site at
http://www.taylorandfrancis.com

and the CRC Press Web site at
http://www.crcpress.com

Contents

Section II   Extensions

Section III   Applications

Section IV   Epilogue

Series Preface

The *Frontiers in Neuroengineering* series presents the insights of experts on emerging experimental techniques and theoretical concepts that are or will be at the vanguard of neuroscience. Books in the series cover topics ranging from electrode design methods for neural ensemble recordings in behaving animals to biological sensors. The series also covers new and exciting multidisciplinary areas of brain research, such as computational neuroscience and neurengineering, and describes breakthroughs in biomedical engineering. The goal is for this series to be the reference that every neuroscientist uses to become acquainted with new advances in brain research.

Each book is edited by an expert and consists of chapters written by leaders in a particular field. The books are richly illustrated and contain comprehensive bibliographies. Chapters provide substantial background material relevant to the particular subjects.

We hope that as the volumes become available, our efforts as well as those of the publisher, the book editors, and the individual authors will contribute to the further development of brain research. The extent to which we achieve this goal will be determined by the utility of these books.

Sidney A. Simon, PhD
Series Editor

Preface

Editing *Methods in Brain Connectivity Inference through Multivariate Time Series Analysis* gave us the chance to reflect on this new and exciting field based on the exploitation of observational brain data. The field is growing at a fast pace with new data/signal processing proposals emerging so often as to make it difficult to be fully up to date. We therefore opted to offer a consolidated panorama of data-driven (as opposed to model-driven) methods with their theoretical basis allied to computational tools to provide the reader with immediate hands-on experience. Alas, not all of our coauthors could supply those tools, something understandable since even good software is frequently hard to employ and often needs to be written for user-friendliness—an all too specialized skill in itself. Nevertheless, we believe that both the data and the supplied softwares do provide a good starting point.

This area depends heavily on time-series analysis tools. Some of them were inspired and born to address specific brain connectivity problems but translate immediately back into time-series analysis and allied application areas. The bridge back, however, has not yet been crossed and we hope that the current tome also proves a helpful incentive to revisit time-series issues.

We would like to thank the series editor for the freedom given to us, and which we hope, we transferred to our chapter authors—all of them original pioneers—whose differences in both approach and opinion expose open problems and throw light onto possible pathways for further research. It should be noted that many chapters contain brand new material, thereby adding value to the book.

Financial support for this endeavor came from FAPESP-CInAPCe Grant 2005/56464-9; CNPq Grants 304404/2009-8 (L.A.B.) and 309381/2012-6 (K.S.); FFM to K.S.

Software and Data CD

Supplemental illustrative material for the book is contained in the form of a CD whose directories are organized by chapter and contain instruction files that provide more details in each case.

Because it refers to more than one chapter, the AsympPDC software package is placed in a separate specific directory; the routines for the examples in Chapters 4 and 7 can be called by their referenced names in the chapters.

The list of chapters possessing supplementary material in the CD can be accessed through the file index.html using a standard browser.

A mirror for the current book material can be found at: `http://www.lcs.poli.usp.br/~baccala/pdc/CRCBrainConnectivity/index.html`.

Editors

Koichi Sameshima is a second-generation hibakusha and native of Minamisatsuma, Kagoshima Prefecture, Japan, who through several hardships graduated in electrical engineering and medicine from the University of São Paulo. He was introduced into the realm of cognitive neuroscience, brain electrophysiology, and time-series analysis during the doctoral and postdoctoral training at the University of São Paulo and the University of California San Francisco, respectively. His research themes revolve around neural plasticity, cognitive function, and information processing aspects of mammalian brain through behavioral, electrophysiological, and computational neuroscience protocols. To functionally characterize collective multichannel neural activity and correlate with animal or human behavior, normal and pathological, he has been pursuing and developing robust and clinically useful methods and measures for brain dynamics staging, brain connectivity inferences, and so on. He holds an associate professorship at the Department of Radiology and Oncology, Faculty of Medicine, University of São Paulo.

Luiz A. Baccalá, after majoring in electrical engineering and physics at the University of São Paulo (1983/1984), he furthered his study on time-series evolution of bacterial resistance to antibiotics in a nosocomial environment and obtained an MSc at the same University (1991). He has since been involved in statistical signal processing and analysis and obtained his PhD from the University of Pennsylvania (1995) by proposing new statistical methods of communication channel identification and equalization. Since his return to his alma mater he has taught courses on applied stochastic processes and advanced graduate level statistical signal processing courses that include wavelet analysis and spectral estimation. His current research interests focus on the investigation of multivariate time-series methods for neural connectivity inference and for problems of inverse source determination using arrays of sensors that include fMRI imaging and multielectrode EEG processing.

Contributors

Laura Astolfi
Department of Computer, Control, and Management Engineering
University of Rome Sapienza
Rome, Italy

Fabio Babiloni
Department of Physiology and Pharmacology
University of Rome Sapienza
Rome, Italy

Luiz A. Baccalá
Department of Telecommunications and Control Engineering
University of São Paulo
São Paulo, Brazil

Katarzyna Blinowska
Department of Biomedical Physics
University of Warsaw
Warsaw, Poland

Aparecida M. Catai
Department of Physiotherapy
Federal University of São Carlos
São Carlos, Brazil

Philip J. A. Dean
The Brain and Behaviour Group
University of Surrey
Guildford, United Kingdom

Luca Faes
Department of Physics and BIOtech
University of Trento
Trento, Italy

Celso Grebogi
Freiburg Institute for Advanced Studies (FRIAS)
University of Freiburg
Freiburg, Germany

Maciej Kamiński
Department of Biomedical Physics
University of Warsaw
Warsaw, Poland

Wei Liao
Center for Cognition and Brain Disorders and the Affiliated Hospital
Hangzhou Normal University
Hangzhou, China

Malenka Mader
Freiburg Center for Data Analysis and Modeling (FDM)
University of Freiburg
Freiburg, Germany

Wolfgang Mader
Freiburg Center for Data Analysis and Modeling (FDM)
University of Freiburg
Freiburg, Germany

Daniele Marinazzo
Department of Data Analysis
University of Ghent
Ghent, Belgium

Nicola Montano
Department of Clinical Science
University of Milan
Milan, Italy

Pedro A. Morettin
Department of Statistics
University of São Paulo
São Paulo, Brazil

Mario Pellicoro
Dipartimento di Fisica
Università degli Studi di Bari and INFN
Bari, Italy

Andrea Plano
Institute of Medical Sciences,
University of Aberdeen
Aberdeen, United Kingdom

Bettina Platt
Institute of Medical Sciences,
University of Aberdeen
Aberdeen, United Kingdom

Alberto Porta
Department of Biomedical Sciences for
  Health
University of Milan
Milan, Italy

Gernot Riedel
Institute of Medical Sciences
University of Aberdeen
Aberdeen, United Kingdom

Koichi Sameshima
Department of Radiology and Oncology
University of São Paulo
São Paulo, Brazil

João Ricardo Sato
Center of Mathematics, Computation
  and Cognition
Federal University of ABC
Santo André, Brazil

Björn Schelter
Institute for Complex Systems and
  Mathematical Biology (SUPA)
King's College
University of Aberdeen
Aberdeen, United Kingdom

Linda Sommerlade
Institute for Complex Systems and
  Mathematical Biology (SUPA)
King's College
University of Aberdeen
Aberdeen, United Kingdom

Sebastiano Stramaglia
Dipartimento di Fisica
Università degli Studi di Bari and INFN
Bari, Italy

Anielle C. M. Takahashi
Department of Physiotherapy
Federal University of São Carlos
São Carlos, Brazil

Daniel Y. Takahashi
Psychology Department and
  Neuroscience Institute
Princeton University
Princeton, New Jersey

Marco Thiel
Institute for Complex Systems and
  Mathematical Biology (SUPA)
King's College
University of Aberdeen
Aberdeen, United Kingdom

Jens Timmer
Freiburg Center for Data Analysis and
  Modeling (FDM)
University of Freiburg
Freiburg, Germany

Gilson Vieira
Functional Neuroimaging Laboratory
  (LIM44)
Institute of Radiology Hospital das
  Clínicas (FMUSP)
São Paulo, Brazil

1 Brain Connectivity
An Overview

Luiz A. Baccalá and Koichi Sameshima

CONTENTS

1.1 INTRODUCTION

The Egyptian days of altogether dismissing the brain as an important organ are long gone. Yet the path to explicit brain connectivity consideration has been a very slow one for even nowadays many research initiatives still center around nineteenth century phrenologically inspired ideas consisting of mainly spotting active brain sites under different stimuli or identifying specific brain structures where cognitive functions can be uncovered under stimulation, inhibition, or lesion protocols.

Acknowledging the central role that must be associated with studying the interplay between neurally active regions has been surprisingly slow, perhaps due to the lack of adequate study tools, with the first reliable results hardly older than perhaps one and a half decades. The intent of this volume is to collect and present a core subset of connectivity inference techniques with an emphasis on their reliability as tools, some of which have been included here for use, testing, and familiarization.

It must be acknowledged that this is a difficult field, not only because so much of it is still so very new but also because many of its conceptual foundations are as yet not part of the standard training of many neuroscientists, most of whom are of a medical or biological persuasion with only the very recent aggregation of engineers, physicists, and mathematicians who may be more readily at ease with the subject matter of many of the chapters that follow.

The book aims at contributing to improve this situation by fostering a wider understanding of the current connectivity issues together with providing a clear picture of the benefits these techniques offer while at the same time stressing the current technical limitations.

Where connectivity analysis is concerned, by far the most widespread analysis tool is *correlation* possibly not only because it is the oldest approach but also because it is simple to understand, being part of even the most basic undergraduate courses.

1

Because neuroelectric signals such as the EEG (electroencephalogram) are readily interpretable physiologically in terms of typical rhythms (alpha, beta, etc.; see Başar, 2004), *spectral coherence*, the frequency domain counterpart of correlation becomes important as well.

Despite the ubiquity and relative simplicity of the latter concepts, they turn out to be largely inadequate in many cases and often provide skewed connectivity pictures as shown in Baccalá and Sameshima (2001). In large measure, much of our goal will be achieved if readers realize that computing correlation coefficients between pairs of dynamically evolving variables is not enough and can be advantageously replaced by the current tools which are far more effective in describing interactions.

What are the central questions that need to be addressed when talking about brain connectivity?

The first that comes to mind is: when can we ascertain that regions are truly interacting? This may be termed the *connectivity detection* problem.

Naturally, one may claim that some form of link exists if their measured activities are correlated—this has been termed "functional connectivity." The activity between two regions may covary even if structures are anatomically unconnected—their correlation may be due to the activity of some other hidden (unmeasured) *driving* structure whose signal, even if only partially, is what is responsible for their communality. To actually be driving the dynamics of others, a structure must be related to the other structures in a physically strong sense. This, however, may happen directly via active anatomic links, or indirectly, if its signal must first be processed in part by intermediary structures. This allows splitting of the nature of connections into the class of *direct* connections and into the class of *indirect* signal pathway connections. This topic is further elaborated in Chapter 13.

As a complicating factor, it is often the case that it is not just driving (i.e., feedforward) alone that matters, but feedback as well whose occurrence needs proper characterization. Hence, if two structures are up for study, their relationship may be *directed* (unidirectional) when one structure is driving the other or bidirectional when feedback is present. As opposed to the techniques addressed herein, pure correlation/coherence methods are unsuited to allow this distinction.

The proper characterization of neural signal driving requires placing of the connectivity problem against an appropriate background: that of time-series analysis where a multivariate setting is required if one is to investigate whether signals can produce effects by traveling through intermediary structures.

The multivariate techniques discussed in this book are centered on evidence-based, that is, data-driven criteria as opposed to structural model-based alternatives, as is the case of DCM (dynamic causal modeling) proposed by Friston et al. (2003) (or even SEM—structural equation models—Loehlin, 2004; Kline, 2005), which require *a priori* guesses as to the underlying interaction mechanisms and for which many algorithms exist to select the most likely structure among those previously hypothesized ones. However, because of the ubiquity of the latter methods in fMRI, they are briefly overviewed in Chapter 11.

As will become readily apparent from the material in this book, a considerable amount of reliable material is becoming available that allows consistent examination of the data-driven case, which has the advantage of fewer presumptions on the researcher's part. We chose not to contrast data-driven against model-driven methodologies since the latter are already covered by a great many publications (Penny et al., 2006; Grefkes et al., 2008; David, 2009; Friston, 2009; Roebroeck et al., 2009) and because we believe both approaches are complementary rather than mutually exclusive. Furthermore, balanced comparative appraisal is perhaps best performed by researchers unattached to either view.

After addressing/solving the detection problem, the next clear task is to characterize connectivity both dynamically and as to its strength (size effect) opening a vast array of possibilities: Is the interaction linear? Is it dependent on the signal frequency content? How quickly does it propagate? How reliable are the answers as brain signal characteristics change while animals exhibit their behavior?

This subject area seeks answers to these questions, although they are still provisional ones in many cases.

1.2 CONNECTIVITY AND CAUSALITY

When are two structures connected? When can their connection be detected? The analogy that comes to mind is that of arteries and veins—they connect organs if blood can flow between them and thereby modify the circumstances of the regions within their reach, either to support a state that would deteriorate if left to itself or to modify its natural condition if no connection existed.

The same applies to electrical wires, whose presence is relevant when the connection is active and the charge is flowing. When electric potentials become identical and charges no longer flow, they are like inactive nerves and neither sustain a state nor change it; they do not represent an active connection and therefore may be removed without changing anything.

These scenarios are familiar to experimenters who have always dealt with it by manipulating how parts of their experimental apparatuses couple and by looking at what happens when equilibrium conditions are disturbed, for instance, arteries can be clamped, wires cut, and so on.

When connections are active and relevant, it is through them that system parts impose the consequences of their activity, which can be seen as *causes* for changing the behavior of other affected system parts leading quite naturally to the idea that *connectivity* and *causality* go hand in hand.

The luxury of invasive manipulation is not always at the experimenter's disposal either due to technical limitations for introducing controlled constraints or because of dealing with *anima nobilis*. In this case, dynamic observation alone is the guide. In modern neuroscience, a number of data-acquisition methods comprising EEG, MEG, and fMRI allow neural activity to be observed without disturbing the system. To some extent, even some biologically invasive techniques like field potential and multi-unit acquisition are known not to provoke significant disruption of normal tissue operation and thus provide data that are essentially observational rather than interventional.

Consequently, when the possibilities of intervention are restricted either intentionally or otherwise, only observational data are available in the form of time-series observations.

There are many good introductions to multivariate time-series analysis as in Lütkepohl (2005). Prediction is one of the main goals of time-series analysis, that is, the study of how to best use recent past observations to provide the best possible guess as to what is going to happen next. Practical uses go from weather forecasts to stock market investment decisions.

For the sake of connectivity appraisal through an exclusively noninterventional observation, one can safely say that the introduction of a notion that became known as Granger causality (G-causality) is of central importance (Granger, 1969).

Granger argued that some form of causality from a time series $x(n)$ onto another time series $y(n)$ may be at play if knowledge of $x(n)$'s past behavior proves helpful in predicting $y(n)$. In fact, all one can safely say is that $x(n)$'s past contains information that is helpful in predicting $y(n)$ and this may be true even in the absence of anatomical links—all that is required is that $y(n)$'s pending fate should become more predictable. An example is when a given perturbation manifests itself first at $x(n)$ and later at $y(n)$ as when some other unmeasured structure affects both the latter (mutually unconnected) structures but whose signal reaches $x(n)$ first. It is this cautionary proviso that makes one term this kind of relationship G-causality, as opposed to the usual general notion of causality, which cannot dispense intervention of some kind for its confirmation.

When the possibility of direct intervention is ruled out, this kind of time-series prediction analysis is the best one can hope for. Naturally, if actual physical means of interaction do exist as when nerve projections are present, it is not unreasonable to accept that true causal influence is at play. G-causality can then become a systematic means of generating physiologically plausible hypotheses regarding the neural phenomenon under the available observations.

One of the interesting properties of G-causality is that it is inherently directional. If

$$G: \ x(n) \rightarrow y(n) \tag{1.1}$$

holds, this does not imply that

$$G: \ y(n) \rightarrow x(n) \tag{1.2}$$

should hold as well.

This is in marked contrast to correlation:

$$C: \ y(n) \leftrightarrow x(n)$$

which is symmetric and precludes signal directionality inference. It is this distinction that makes room for the notion of *directed* connectivity mentioned in Section 1.1. Feedback is detected when both Equations 1.1 and 1.2 hold simultaneously.

Pairwise Granger causality introduces the possibility of detecting signal flow directionality, and this alone is a huge leap forward from correlation studies alone.

However, it is possible to show (Baccalá and Sameshima, 2001) that false causal connectivity structures may result from considering just pairwise G-causality; this calls for the simultaneous consideration of as many relevant observed time series as possible. Observing many time series, however, has the undesirable consequence of increasing *inferential complexity*, which negatively impacts statistical inference reliability. This remains a problem that is largely unsolved in practice, specially in signal imaging modalities like MEG and fMRI whose high spatial resolution produces a large number of time series, usually of short physiologically interesting duration for fully reliable joint causality statistical consideration.

Despite the latter difficulties, joint consideration of up to 27 simultaneous time series is feasible in some cases, and this leads to reasonable cross patient estimates of epileptic foci during ictal EEG episodes as described in Baccalá et al. (2004). See also Example 7.7.

As the conceptual common denominator of the techniques presented in this book, G-causality deserves some additional comments before we move on to an overview of the contributions herein.

Since G-causality is based on the idea of prediction, its inference is intimately tied to how prediction is to be performed.

To be able to predict, one needs to define how prediction is to be carried out: namely, by linear or nonlinear methods. One must also agree on what constitutes the available past information (how far into the past to look) whose inclusion has the effect of improving prediction over not considering this information under the predictive method of choice. G-causality detection therefore depends on many analysis choices.

In many application areas, linear models for relating observations are the predictive models of choice. A number of mostly practical reasons explain this: linear models are reliable and easy to compute. This is specially true for autoregressive models (see Lütkepohl, 2005; and Chapter 3); they lead to unique solutions that are optimal when the data are Gaussian. In some cases of non-Gaussian data, nonlinear models may be more appropriate but they are far more difficult to fit and to diagnose as to prediction quality. Nonlinear models are nonetheless interesting and important for their revelatory powers with regard to the physical details of the connections but constitute a second-order step in analysis refinement that often suffers from further nontrivial inferential complexity impairments. Model-free (nonparametric) prediction methods are also possible but also often require massive data quantities for the accurate estimation of conditional probabilities based on relevant information sets. This often relegates these methods to the realm of impracticality, however theoretically appealing they ultimately are. In this book, we concentrated on methods of immediate applicability rather than on the examination of general abstract strategies. We believe the reader should first acquire practical tools to start working with these ideas as soon as possible.

1.3 CHAPTERS OVERVIEW

The book is composed of three main sections. Section I covers the fundamentals. Chapter 2 reviews the pioneering contribution of Kamiński and Blinowska and their introduction of *directed transfer function* (DTF) as a frequency domain descriptor of multivariate neural connectivity. This is followed by an in-depth discussion of the basic role played by multivariate autoregressive modeling, its issues, and its time-domain interpretation in terms of time-domain Granger causality (Chapter 3). Chapter 3 also examines the important subject of model diagnostics, which is key for connectivity inference validity.

The reappraisal of the connectivity issue in terms of direct versus indirect connections is discussed in Chapter 4 through the notion of *partial directed coherence*.

A more abstract take is employed to examine the latter quantities from the standpoint of information flow leading to the notion of information PDC and information DTF (Chapter 5). Also along abstract lines, Chapter 6 examines the issue of instantaneous causality—that is, the description and meaning of the residual correlation between the prediction errors that results even after successful multivariate time-series modeling.

The last installment of the fundamentals is represented by Chapter 7, which discusses the statistical behavior of PDC and some of its variations and thereby opens the possibility of rigorous connectivity testing in terms of both the *detection* problem and the connectivity *quantification* problem.

An overview of this book's core theoretical subject may be appreciated through a classification of some of its fundamental papers in Figure 1.1. The figure is organized into four plane regions—the top part representing causality descriptions associated with pairs of time series followed by the bottom part which sums up the strictly multivariate time-series considerations. The left half of the plane states those core time-domain analysis contributions whose frequency domain counterparts are displayed on the right side of the plane. Most of the relevant multivariate frequency domain historical development is further elaborated in Chapters 2 and 4, whereas a generalized summary of the inferential properties (marked by dashed boundaries in Figure 1.1) is examined in Chapter 7 leaving information theory considerations to Chapter 5. Note that early contributions to the subject come from the Japanese school represented by Akaike (1968) who already made feedback considerations even before Sims (1972) re-examined Granger's theory (Granger, 1969) in terms of feedback, leading to the first time-domain bivariate statistical inference efforts (Geweke, 1979) that culminated in Lütkepohl's monumental 1993 work (see also Lütkepohl, 2005). Also outstanding are Geweke's works of 1982 and 1984 at the interface between time and frequency domain descriptions. To some measure, arrows portray the conceptual dependence between the contributions without necessarily indicating whether the authors were fully aware of one another's works at the time of their writing.

Section II is composed of recent contributions that point to some pathways for future development that include tackling nonlinear interactions (Chapter 8) that usual multivariate linear autoregressive models cannot capture and the problem of signal nonstationarity through time-variant models (Chapter 9).

Section III covers applications. In Chapter 10, connectivity models are used in conjunction with EEG dipole source reconstruction algorithms. This is followed by fMRI applications in Chapter 11. The effectiveness of G-causality appraisals is investigated while considering the relationship of very general multimodal biological variables (Chapter 12).

The book concludes with an epilogue in Chapter 13 which takes a look back at the contributions, points to some future developments, but most importantly proposes

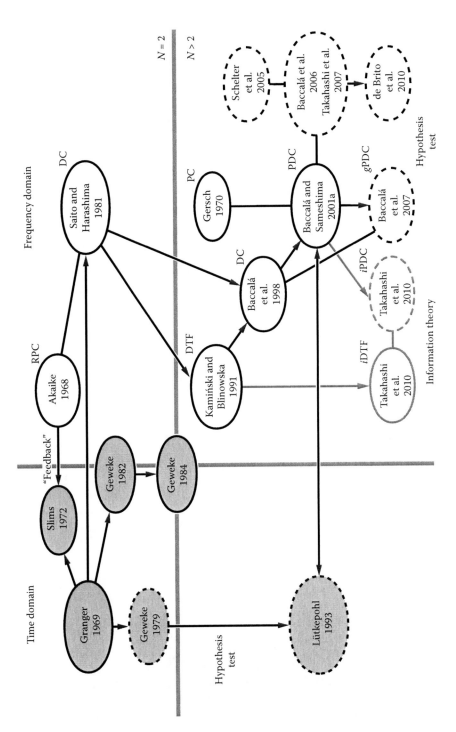

FIGURE 1.1 Synoptic diagram of the core contributions to this book's subject.

a new framework for discussing connectivity by replacing the popular notions of functional/effective connectivity.

1.4   LARGER AUDIENCE

Even though computational neuroscience has been the birthplace of the ideas presented in this book, its techniques are of much wider applicability such as climate characterization, see Example 7.6, and we hope time-series practitioners may profit from the application of the present ideas and techniques.

The enclosed software, especially the AsympPDC package, has been written for code clarity rather than efficiency and is included in the CD that accompanies the book.

REFERENCES

Akaike, H. 1968. On the use of a linear model for the identification of feedback systems. *Ann. Inst. Stat. Math. 20*: 425–440.

Baccalá, L. A., K. Sameshima, G. Ballester et al. 1998. Studying the interaction between brain structures via directed coherence and Granger causality. *Appl. Sig. Process. 5*: 40–48.

Baccalá, L. A. and K. Sameshima. 2001a. Partial directed coherence: A new concept in neural structure determination. *Biol. Cybern. 84*: 463–474.

Baccalá, L. A. and K. Sameshima. 2001b. Overcoming the limitations of correlation analysis for many simultaneously processed neural structures. *Prog. Brain Res. 130* (Advances in Neural Population Coding): 33–47.

Baccalá, L. A., M. Alvarenga, K. Sameshima et al. 2004. Graph theoretical characterization and tracking of the effective neural connectivity during episodes of mesial temporal epileptic seizure. *J. Integ. Neurosci. 3*: 379–395.

Baccalá, L. A., D. Y. Takahashi, and K. Sameshima. 2006. Computer intensive testing for the influence between time series. In *Handbook of Time Series Analysis*, R. Schelter, M. Winterhalter, and J. Timmer (Eds.), 411–436. Weinheim: Wiley-VCH.

Baccalá, L. A., D. Y. Takahashi, and K. Sameshima. 2007. Generalized partial directed coherence. In *15th International Conference on Digital Signal Processing* (DSP2007), 163–166, Cardiff, Wales, UK.

Başar, E. 2004. *Memory and Brain Dynamics: Oscillations Integrating Attention, Perception, Learning, and Memory*. Boca Raton, FL: CRC Press.

David, O. 2009. fMRI connectivity, meaning and empiricism: Comments on: Roebroeck et al. The identification of interacting networks in the brain using fMRI: Model selection, causality and deconvolution. *NeuroImage 58*: 306–309.

de Brito, C., L. A. Baccalá, D. Y. Takahashi et al. 2010. Asymptotic behavior of generalized partial directed coherence. In *Engineering in Medicine and Biology Society (EMBC), 2010 Annual International Conference of the IEEE*, 1718–1721, Buenos Aires, Argentina.

Friston, K., L. Harrison, and W. Penny. 2003. Dynamic causal modelling. *NeuroImage 19*: 1273–1302.

Friston, K. 2009. Causal modelling and brain connectivity in functional magnetic resonance imaging. *PLoS Biol. 7*, e1000033.

Gersch, W. 1970. Epileptic focus location: Spectral analysis method. *Science 169*: 701–702.

Geweke, J. 1979. Testing the exogeneity specification in the complete dynamic simultaneous equation model. *J. Econometr. 7*: 163–185.

Geweke, J. 1982. Measurement of linear dependence and feedback between multiple time series. *J. Am. Stat. Assoc. 77*: 304–324.

Geweke, J. 1984. Measures of conditional linear dependence and feedback between time series. *J. Am. Stat. Assoc. 79*: 907–915.

Granger, C. W. J. 1969. Investigating causal relations by econometric models and cross-spectral methods. *Econometrica 37*: 424–438.

Grefkes C., S. B. Eickhoff, D. A. Nowak et al. 2008. Dynamic intra- and interhemispheric interactions during unilateral and bilateral hand movements assessed with fMRI and DCM. *NeuroImage 41*: 1273–1302.

Kamiński, M. and K. J. Blinowska. 1991. A new method of the description of the information flow in brain structures. *Biol. Cybern. 65*: 203–210.

Kline, R. 2005. *Principles and Practice of Structural Equation Modelling* (2nd ed.). New York: Guilford Press.

Loehlin, J. 2004. *Latent Variable Models: An Introduction to Factor, Path, and Structural Equation Analysis* (4th ed.). Mahwah: Lawrence Erlbaum.

Lütkepohl, H. 1993. *Introduction to Multiple Time Series Analysis*. New York: Springer.

Lütkepohl, H. 2005. *New Introduction to Multiple Time Series Analysis*. New York: Springer.

Penny, W. D., K. J. Friston, J. T. Ashburner et al. (Eds.). 2006. *Statistical Parametric Mapping: The Analysis of Functional Brain Images*. London: Academic Press.

Roebroeck A., E. Formisano, and R. Goebel. 2009. The identification of interacting networks in the brain using fMRI: Model selection, causality and deconvolution. *NeuroImage 58*: 296–302.

Saito, Y. and H. Harashima. 1981. Tracking of information within multichannel record: Causal analysis in EEG. In *Recent Advances in EEG and EMG Data Processing*, 133–146. Amsterdam: Elsevier.

Schelter, B., M. Winterhalder, M. Eichler et al. 2005. Testing for directed influences among neural signals using partial directed coherence. *J. Neurosci. Methods 152*: 210–219.

Sims, C. A. 1972. Money, income, and causality. *Am. Econ. Rev. 62*: 540–552.

Takahashi, D. Y., L. A. Baccalá, and K. Sameshima. 2010. Information theoretic interpretation of frequency domain connectivity measures. *Biol. Cybern. 103*: 463–469.

Section I

Fundamental Theory

2 Directed Transfer Function
A Pioneering Concept in Connectivity Analysis

Maciej Kamiński and Katarzyna Blinowska

CONTENTS

2.1 INTRODUCTION

Over the last half century, neuroscience largely concentrated itself on the problems of localization of brain centers connected with recognition of a given stimulus or with a given behavior. Besides anatomical studies, different imaging methods based mainly on the metabolism changes in brain were developed. The problem of identification of the brain activity sources was approached by means of varied inverse problem solutions. However, understanding brain information processing requires identification of the causal relations between active centers. To this aim, methods for estimation of causal influence that one system exerts another were developed. During the last 10 years, the problem of functional connectivity became a focus of attention in brain research. Here, we describe the development of methods, which became valuable tools for estimation of the transmission between brain structures and, in particular, for the assessment of dynamic information processing by the brain.

2.2 PARAMETRIC APPROACH

In spectral analysis, we consider the properties of the signals in the frequency domain and the basic measure is the power spectral density (or power spectrum) describing the frequency content of a signal. There are various approaches for estimation of the power spectrum. In general, we can divide these methods into two groups: nonparametric and parametric. Nonparametric methods derive spectral quantities directly from the signal values. The Fourier transform is a nonparametric method that is very popular in every field of data analysis. This simple and straightforward implementation method makes one assumption about the signal, namely, that an infinite or periodic signal can be represented by a certain number of sinusoids. There exists a fast and effective algorithm for its estimation called FFT (fast Fourier transform).

Parametric spectral estimation methods assume a model of data generation underlying the observed time series. The process of analysis starts from fitting the assumed model to the signal. After that, we derive all necessary quantities from the fitted model parameters, not from the signal itself. Such an approach has certain advantages over nonparametric methods: spectra obtained by parametric methods do not have the windowing problem, common in Fourier's analysis—the Fourier transform is defined for infinite signals while every experimental data record is finite. So, in fact, we must calculate the transform of the hypothetical infinite signal multiplied by a finite window of observation. The transformation to the frequency domain is then a transform of the signal convolved with the transform of the window function, which leads to the distortion of the spectral estimate. The parametric approach does not require window functions; it assumes that we observe a part of an infinite signal described by the same model. As a result, we get smoother spectra with better statistical properties, especially for shorter signals. Moreover, the parameters of the data generation model allow the construction of new estimators, including multivariate causal estimators.

The parametric method requires the choice of a proper model of the data generation. The most popular linear model in biomedical data analysis is the autoregressive model (AR). The properties of such a model agree well with the characteristics of typical EEG data and many other biological signals as well. The spectrum of an AR model has a theoretical form of a fraction with a polynomial (constructed from the model parameters) in the denominator. Functions of such type describe spectra with several maxima. The impulse response function of the AR model consists of a sum of damped sinusoids. These components may be interpreted as "rhythms" of the EEG, as was shown in Blinowska and Franaszczuk (1989). Since we can analytically calculate parameters of the impulse response function, the amplitude and frequency position of every rhythmical component can be determined without calculation of the spectrum. The method of the estimation of the parameters of rhythmical components was called FAD (frequency, amplitude, damping) (Franaszczuk and Blinowska, 1985).

2.3 HISTORICAL ATTEMPTS

The earliest attempts at describing the influence of one variable on another came from social sciences and economics, starting from the 1950s. Because the computer was not a typical researcher's tool (as it is today), the first measures had a more descriptive rather than quantitative character. It is worth noting that the first causal models did not include temporal relations between variables. Assuming as a null hypothesis independent realizations of separate random variables, hypothetical directions of influence were assumed and then tested. Such an approach evolved into a method that is today called "structural equations modeling," which is still used in data analysis (McIntosh and Gonzalez-Lima, 1994).

The temporal ordering of data was included in consideration of causal relations by Wiener (1956). The idea of causal relations relies on the predictability of time series. Typically, if the information contained in one time series could help in prediction of the other time-series values, the first one could be called causal for the second one. That purely statistical and descriptive concept was adapted into the data analysis field by Granger (1969). He introduced the causality measure founded on the linear systems theory, which later led to the development of different estimators of causality working in the time or the frequency domain, based on the Granger causality concept.

In the 1970s and 1980s, an increasing number of publications appeared, discussing various types of measures, and designed to describe relations among signals. Especially useful were measures, that operated in the frequency domain. From that time, we should mention Geweke's (1982; 1984) works on decomposition of multivariate autoregressive process, studies of feedbacks in linear systems by Gevers and Anderson (1981) and by Caines and Chan (1975).

Practically, all the measures proposed for estimating the relationship between signals were defined for two signals only. That imposes serious limitations on the analysis of multichannel data and may be a source of pitfalls—that issue will be discussed later.

Our idea of defining a frequency-dependent causality measure, which would be easy to apply in biomedical data analysis and would have a truly multichannel character, has its roots in the paper of Saito and Harashima (1981). The paper introduces (among others) directed coherence. In their two-channel model, besides the noise feeding two signals of the system, there was also a third noise signal common to both channels. The efforts to extend the proposed formalism to a larger number of channels were not successful, since the number of parameters increased rapidly with the number of channels, which made it impossible to solve the resulting equations. Another approach, which happened to be promising in multivariate data estimation, has already been mentioned: autoregressive modeling.

Linear systems and specific models such as autoregressive (AR), moving average (MA), or autoregressive-moving average (ARMA) played a substantial role in the causal analysis considerations. The subject got a great deal of attention in Japan. Many papers dating back to the late 1960s (e.g., Akaike, 1968, 1974; Wang et al., 1992; and others) describe properties of such models and their applicability in multivariate data analysis. Somehow, the discoveries made by the Japanese scientists, due to the lack of quick bibliography search methods, had to be rediscovered by other researchers years later. The AR model applications were already present in the literature, as a tool to estimate spectra of a single time series, primarily in the field of geophysical applications. It was demonstrated by Ulrych and Bishop (1975) that power spectrum obtained from the AR model is identical to that obtained by maximizing the entropy of the process. The maximum entropy estimate takes into account the known constraints and is maximally noncommittal with regard to unavailable information. In consequence, AR spectral estimates have favorable statistical properties.

2.4   MULTIVARIATE DATA

A multivariate data set is an ensemble of simultaneously recorded signals (channels). Each signal is a separate time series, but since the recording was made synchronously for all the signals, we expect the set to contain more information than was stored in each channel separately. That additional information is contained in relations between signals. The measures describing such relations are called *cross-quantities* (by contrast to *auto-quantities* describing separate properties of every signal). The system generating the signals is described by information content including auto- and cross-quantities.

The most popular cross-quantities are without any doubt, correlation and covariance (correlation computed after subtraction of the mean value from the signals). Correlation and covariance are measures describing similarities of both signals, which are consistent in the time domain. Very often, frequency-dependent measures are of interest; after transforming the correlation to the frequency domain, one gets a corresponding measure called coherence. The coherence describes similarities of signals' frequency components which are consistent in phase.

Although the phase of coherence in principle should tell us about the direction of the influence between signals (the consistent phase shift translates into consistent time difference in the time domain, which denotes precedence of influence), in practice the phase of coherence may give ambiguous results, as its values are defined (and the direction changes) modulo 2π. Also, reciprocal connections between signals cannot be found. Moreover, the structure of real biomedical signals is complex, making phase values change rapidly and lead to results that are difficult to interpret.

Another problem with multivariate data sets arises for systems consisting of more than two channels. In such cases, more complicated signal relations may take place. Specifically, one signal may be a source of activity for many other signals (common feeding problem) or the activity may be transmitted from source to the destination channel directly or through other channels. In such situations, bivariate measures likely give misleading results, which will be demonstrated below.

Owing to the complex nature of interrelations within a multichannel set of signals, the appropriate analysis tool must use all the information contained in the set to estimate the directions and strength of influences. Such a tool may be defined in the framework of the parametric modeling approach.

2.5   AUTOREGRESSIVE MODEL

The multivariate AR model (MVAR) is defined for a k-channel set (a set of k simultaneously observed signals) as

$$X(t) = \sum_{j=1}^{p} A(j)X(t-j) + E(t), \qquad (2.1)$$

where: $X(t) = [X_1(t), X_2(t), ..., X_k(t)]^T$ is a vector of k signals' values at a time t, $E(t) = [E_1(t), E_2(t), ..., E_k(t)]^T$ is a vector of k white noises' values at a time t, and

$$\mathbf{A}(j) = \begin{pmatrix} A_{11}(j) & \cdots & A_{1k}(j) \\ \vdots & \ddots & \vdots \\ A_{k1}(j) & \cdots & A_{kk}(j) \end{pmatrix} \text{ for } j = 1, \ldots, p \text{ are the model parameters, which}$$

we estimate from the data. Let $\mathbf{A}(0)$ equal the identity matrix.

The value p is called the model order.

Model fitting quality depends on the proper model order selection. Selecting too low orders may give us spectra lacking the necessary details, while too big orders tend to create spurious maxima in the spectrum. The model order may be found from criteria derived from the information theory (McQuarrie and Tsai, 1998). Several criteria have been proposed, e.g.:

1. Akaike's information criterion:

$$\text{AIC}(p) = \ln[\det(\mathbf{V})] + 2pk^2/N$$

2. Hannan–Quinn's criterion:

$$\text{HQ}(p) = \ln[\det(\mathbf{V})] + 2\ln(\ln(N))pk^2/N$$

3. Bayesian–Schwartz's criterion:

$$\text{SC}(p) = \ln[\det(\mathbf{V})] + \ln(N)pk^2/N,$$

where $\mathbf{V}$ is the noise $E(t)$ covariance matrix.

Typically, the first term of a criterion depends on the estimated residual variance $\mathbf{V}(p)$ for a given p, and the second term is a function of model order p, number of channels k and number of data points N. The model order has to be a compromise between the accuracy of the fitting (first term) and the excessive number of parameters, which worsens the statistical properties of the estimate. The AIC criterion is the one that is mostly used. Different criteria give similar results for the signals, which are reasonably compatible with the model.

There are several algorithms of model parameters estimation, differing in capability of detecting specific characteristics of the spectrum (like narrow band sinusoidal-like components) or stability for shorter data segments. All these issues are discussed in detail in signal analysis handbooks e.g.: Jenkins (1968); Priestley (1983); Marple Jr. (1987); Kay (1988); Lütkepohl (1993); Box et al. (2008); Hamilton (1994); Schneider and Neumaier (2001); Schelter et al. (2006).

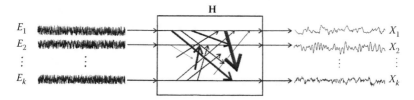

FIGURE 2.1 The AR model as a linear filter with transfer matrix $\mathbf{H}$, noises E on input, and signals X on output. (Reprinted from Kamiński, M. and H. Liang. 2005. *Crit. Rev. Biomed. Eng. 33*: 347–430.)

The AR model Equation 2.1 may be transferred to the frequency domain, obtaining:

$$\mathbf{E}(f) = \mathbf{A}(f)\mathbf{X}(f),$$

$$\mathbf{X}(f) = \mathbf{A}^{-1}(f)\mathbf{E}(f) = \mathbf{H}(f)\mathbf{E}(f),$$

$$\mathbf{H}(f) = \left(\sum_{m=0}^{p} \mathbf{A}(m) \exp(-2\pi imf\,\Delta t)\right)^{-1}. \tag{2.2}$$

That relation has a form of a linear filter operating on white-noise inputs E_i and producing signals X_i as its output (Figure 2.1).

The power spectrum can be calculated directly from the formula:

$$\mathbf{S}(f) = \mathbf{X}(f)\mathbf{X}^*(f) = \mathbf{H}(f)\mathbf{E}(f)\mathbf{E}^*(f)\mathbf{H}^*(f) = \mathbf{H}(f)\mathbf{V}\mathbf{H}^*(f). \tag{2.3}$$

Note that in the above relation we obtain the whole power spectral matrix with auto-spectra on the diagonal and cross-spectra off the diagonal. From the cross-spectra, it is only one step to calculate the (ordinary) coherence:

$$K_{ij}(f) = \frac{S_{ij}(f)}{\sqrt{S_{ii}(f)S_{jj}(f)}}. \tag{2.4}$$

The elements of the matrix $\mathbf{H}$ contain not only the spectral but also the phase information; hence, they may be used to define the causality measure—the DTF function. Originally, it was proposed in the normalized version (Kamiński and Blinowska, 1991):

$$\gamma_{ij}^2(f) = \frac{\left|\mathbf{H}_{ij}(f)\right|^2}{\sum_{m=1}^{k} |\mathbf{H}_{im}(f)|^2}. \tag{2.5}$$

Such quantity may be interpreted as a ratio between the inflow from channel j to channel i to the sum of all inflows to the channel i. The value of normalized DTF is in [0, 1] range: 0 means no influence, 1 means the maximal influence.

Sometimes, we are more interested in detection of a transmission (not the ratio of inflows). For that purpose, the nonnormalized DTF was proposed. It has a form of

$$\theta_{ij}^2(f) = \left| \mathbf{H}_{ij}(f) \right|^2. \qquad (2.6)$$

The nonnormalized DTF is not bound between 0 and 1. Its value is related to the coupling strength between channels, as was demonstrated in Kamiński et al. (2001). Some other versions of DTF functions were also proposed in the literature (Korzeniewska et al., 2003).

2.6 PARTIAL CAUSAL MEASURES, PDC AND DDTF

In multivariate systems with more than two channels, there are many possible patterns of transmission between channels. For instance, a signal can be transmitted from one channel to another, directly or indirectly, through one or more channels. The DTF function detects the presence of the source signal in the destination channel not only for direct flows, but also for cascade flows. In order to distinguish between direct and indirect transmissions, partial measures have to be applied.

PDC (partial directed coherence) was proposed by Baccalá and Sameshima (2001) as a measure showing in the frequency domain direct causal relations only (see also Chapter 4 for further details). It is defined using the frequency domain transformed model coefficients $\mathbf{A}(f)$, not the elements of $\mathbf{H}(f) = \mathbf{A}^{-1}(f)$.

$$P_{ij}(f) = \frac{A_{ij}(f)}{\sqrt{\mathbf{a}_j^*(f)\mathbf{a}_j(f)}}. \qquad (2.7)$$

This is a normalized quantity describing a ratio between outflow from the channel j to channel i to all the outflows from the channel j. Recently, a modification of PDC was proposed, taking into account the variance of the original signal and introducing in the denominator another normalization factor. The new quantity called gPDC (generalized PDC) is given by Baccalá et al. (2007):

$$\pi_{ij}(f) = \frac{A_{ij}(f)\frac{1}{\sigma_i}}{\sqrt{\sum_{m=1}^{k} \left| A_{mj}(f)\frac{1}{\sigma_m} \right|^2}}. \qquad (2.8)$$

The PDC function will be discussed in detail in Chapter 4.

Another partial measure of connectivity, which detects only direct influences, was proposed in Korzeniewska et al. (2003). The direct directed transfer function (dDTF) is a function combining partial coherences and a modification of DTF into one quantity (χ):

$$\chi_{ij}^2(f) = F_{ij}^2(f)C_{ij}^2(f), \tag{2.9}$$

$$F_{ij}^2(f) = \frac{\left|\mathbf{H}_{ij}(f)\right|^2}{\displaystyle\sum_f \sum_{m=1}^{k} |\mathbf{H}_{im}(f)|^2}. \tag{2.10}$$

The value of this function is nonzero only when there is a direct transmission from channel j to channel i. The dDTF was introduced to identify unequivocally the direct connections. It is important to apply dDTF for the electrodes placed in the brain structures. It was introduced explicitly to evaluate the propagation between the brain structures of the behaving animal. In case of the signals registered from the scalp, it is more appropriate to use DTF as was demonstrated in Kuś et al. (2004). The dDTF applied for scalp EEG shows propagation along anatomical connections; however, this information is not of primary importance. It is important to observe which brain sites are communicating and exchanging information.

2.7   GRANGER CAUSALITY AND ITS RELATION TO DTF

The Granger causality, as has already been mentioned, is based on the prediction of time series. A time series X may be predicted by means of its previous values:

$$X(t) = \sum_{j=1}^{p} A_{11}(j)X(t-j) + E_1(t). \tag{2.11}$$

The error of the prediction E_1 has variance V_1. If we now predict $X(t)$ using previous values of X together with values of another process Y, we get

$$X(t) = \sum_{j=1}^{p} A_{11}'(j)X(t-j) + \sum_{j=0}^{q} A_{12}(j)Y(t-j) + E_2(t). \tag{2.12}$$

Now, the prediction error E_2 has in general a different variance V_2. If $V_2 < V_1$, then the prediction gets better and signal Y can be considered as causal for the signal X.

The measures which are directly based on the comparison of the prediction errors are: measure of linear feedback and Granger causality index (Geweke, 1982; Brovelli et al., 2004). In case of DTF/PDC, the prediction error variances V_1 and V_2 are not directly compared. The approach is as follows: we estimate the model (minimizing the variance of the residual noise as best as we can) for both (all) channels and check the value of the coefficient connecting signals X_j and X_i. If that coefficient is close to zero, we say that there is no influence of signal X_j on signal X_i. Both methods use that approach, namely, in case of PDC as a measure of causality, $A_{ij}(f)$ is used; for DTF, it is an element of the transfer matrix $\mathbf{H}_{ij}(f)$.

2.8 MULTIVARIATE VS. BIVARIATE APPROACH

In case of a system of mutually connected channels of a number larger than 2, different connectivity patterns are possible. Quite often the aim of analysis concerns finding the channel that drives the system. When bivariate connectivity measures (two channels at a time are considered) are used, it may be practically impossible to find the correct pattern of propagation. This point is illustrated in Figure 2.2 by a very simple example. In case of a three-channel model, channels 2 and 3 are driven by channel 1. By means of a bivariate analysis, the propagation is found not only from channel 1, but also from channel 2. In fact, the propagation is found always when there is a delay between signals. Let us assume that we record the signals in different distances from the source. We shall obtain nonzero propagation from all proximal channels to the more distant ones; the pattern of connectivity will be very complicated and will reveal a sink of activity at the most distant channel.

Comparison of bivariate and multivariate methods of directionality estimation was considered in Blinowska et al. (2004), Kuś et al. (2004), Kamiński (2005); therefore, we shall not discuss it here in detail. In the above papers, it was demonstrated by means of different simulations and also in the examples of experimental signals

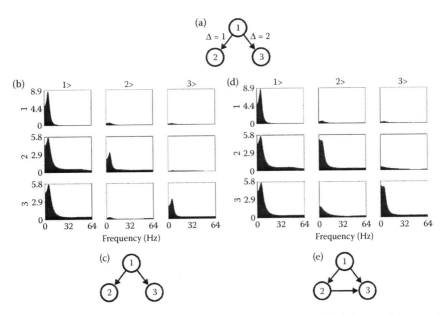

FIGURE 2.2 DTF results for a simulated three-channel system. (a) Scheme of simulated flows. Δ represents the time delay in samples. (b) Multichannel DTF results. Each cell represents a (solid filled) graph of DTF with frequency on the horizontal axis (0–64 Hz) and function value on the vertical axis. Transmission from the channel marked above the relevant column to the channel marked at the left of the relevant row. On the diagonal of the matrix, the power spectra are shown. (c) Scheme of flows deduced from the multichannel DTF. (d) Pairwise DTF results (organization is the same as in b). (e) Scheme of flows deduced from pairwise DTFs. (Reprinted from Kamiński, M. and H. Liang. 2005. *Crit. Rev. Biomed. Eng. 33*: 347–430.)

that bivariate measures can give misleading results. Sometimes, even the reversal of propagation may be found. Brain signals are usually strongly coupled, as was shown, for example, in Kamiński et al. (1997); therefore, the use of fully multivariate measures, that is, considering all channels simultaneously, is a crucial requirement for proper connectivity analysis.

2.9   NONLINEAR DATA AND LINEAR MVAR APPROACH

Very often certain objections are raised against linear modeling efficiency in application to nonlinear systems (with nonlinear relations between signals). Indeed, linear models are able to detect and describe well linear signal relations. However, in case of EEG or other brain signals, the question arises: how often do we really deal with nonlinearities and what impact can it have on the analysis results? The presence of nonlinearities in the signal can be detected by means of the surrogate data tests (Efron and Tibshirani, 1993). In case of EEG time series, the application of these tests did not confirm the nonlinear character of the EEG signal (Achermann et al., 1984; Stam et al., 1999). Also, the comparison of the linear and nonlinear forecasts of EEG and LFP (local field potentials) demonstrated the linear character of brain signals (Blinowska and Malinowski, 1991). The traces of nonlinearity were found only during certain stages of epileptic seizures (Pijn et al., 1991, 1997). The nonlinear methods were widely applied for epileptic EEG; however, the promising results reported were put in doubt after more careful analysis and thorough tests (e.g., Harrison et al., 2005; McSharry et al., 2003 or Lai et al., 2003). In a study of Wendling et al. (2009), coupling in the epileptic brain was estimated by nonlinear regression, phase synchronization, generalized synchronization, and also linear regression. The latter measure outperformed the nonlinear estimators.

It seems that the problem of nonlinear relations is not a crucial issue in EEG analysis. In fact, even in the presence of nonlinearities, we can still get valuable estimates from linear modeling applications. David et al. (2004) showed for simulated nonlinear systems that linear methods may sometimes miss some connections, but they are not likely to give wrong directions of the detected ones. The comparison of nonlinear methods of connectivity (mutual information, phase synchronization, continuity measure) with a linear measure (correlation) was carried out for the nonlinear simulated time series by Netoff et al. (2006). They reported that the linear method performed better in the presence of noise in comparison to all the nonlinear methods tested. The performance of different measures of connectivity applied for the nonlinear time series was also tested by Winterhalder et al. (2005). It was found that DTF and PDC performed quite well when applied to nonlinear signals; however, in case of PDC, a very high model order has to be used.

We present below an example of the application of DTF to localize an epileptogenic focus. The utility of MVAR for the analysis of nonlinear epileptic signals was shown in a case of three patients from the University of Maryland at Baltimore (see Franaszczuk et al., 1994). The patients suffered from pharmacologically intractable epilepsies and were scheduled for surgery. Brain activity was recorded from a 32-contact subdural on-cortex grid array and an 8-contact depth array

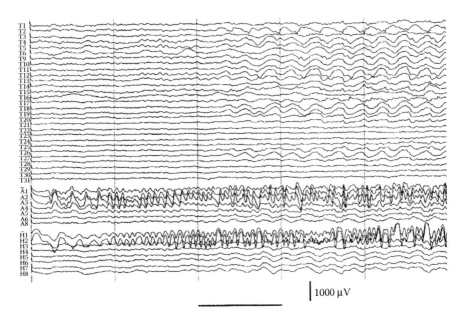

FIGURE 2.3 Ictal recordings from a 32 contact subdural grid over the lateral temporal lobe cortex (T electrodes), and two depth electrodes (A and H electrodes) placed through the grid. The panel illustrates the seizure several seconds after onset from a mesial temporal focus with a predominant 12–14 Hz discharge. The horizontal bar marks the 1 s of the seizure utilized for the DTF analysis (the results are shown in Figure 2.4). (Reprinted from Franaszczuk, P. J., G. K. Bergey, and M. Kamiński. 1994. *Electroenceph. clin. Neurophysiol. 91*: 413–427.)

of electrodes implanted in mesial temporal structures. An example of one of the data records is presented in Figure 2.3. The sampling frequency of the data was 200 Hz. From the original 64-channel recordings, the 16-channel subsets were selected and epochs of 1 s length (or longer for nonstationary parts of the recordings) from different stages of the seizure were chosen.

The analysis was focused on the 1–25 Hz frequency range: the frequency of the epileptic activity varied from patient to patient, but it was within that range. The DTFs were calculated and maxima of their values were found for the identified frequencies of interest (Figure 2.4).

In all the three cases discussed, the DTF method allowed the sources of epileptic activity propagating to other locations to be detected. The accuracy of the localization was confirmed by the fact that after removal of the identified areas by surgical treatment, the ictal activity was stopped.

We can conclude that in case of an EEG recorded from the scalp or from indwelling electrodes, there is hardly a need for application of nonlinear methods. Moreover, nonlinear methods are vulnerable to noise and prone to systematical errors connected with the arbitrary choices one has to make (e.g., concerning lags, embedding dimension, or histogram bins). The nonlinear connectivity measures such as mutual information, generalized synchronization, and nonlinear correlation are limited to

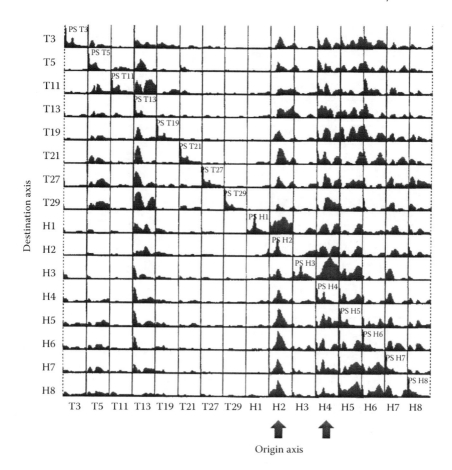

FIGURE 2.4 Directed transfer functions for 16 channels (8 from hippocampal depth electrodes and 8 selected from the subdural grid array) for a 1 s period (see Figure 2.3) near the onset of the seizure. The power spectra for each contact are shown on the diagonal. The predominant peak seen on the EEG is the 12–14 Hz peak; this is the one that is the most prominent in the power spectra from the deep hippocampal contacts. The predominant source of this activity can be found on the origin axis, where H2 and H4 are the greatest contributors (solid arrows). The horizontal scale for each box is 0.5–35 Hz. The vertical scale for DTF is 0–1; the power spectra are normalized to unit power. (Reprinted from Franaszczuk, P. J., G. K. Bergey, and M. Kamiński. 1994. *Electroenceph. clin. Neurophysiol. 91*: 413–427.)

2 channels only, which can lead to substantial errors as was demonstrated in the preceding section.

2.10 NONSTATIONARY DATA

The next logical step in the development of causal multivariate estimators is the analysis of nonstationary data. In the previous applications, we estimated model parameters and spectral functions for long data epochs. The obtained estimates described average properties for the given record of presumably stationary (not changing in time its statistical properties) data. Unfortunately, most of the interesting phenomena are dynamic and we would like to describe their time-varying properties as well. The problem has two general types of solutions: (1) adaptive methods, and (2) short-window method. The first group of methods allow model parameters to change in time and to be adjusted during calculations. They are practically realized utilizing the concept of the Kalman filter—an optimal observer approach for a dynamic process (Gersch, 1987; Benveniste et al., 1990; Gath et al., 1992; Arnold et al., 1998; Hesse et al., 2003; Simon, 2006). In the second type of approach, the data epoch is divided into shorter segments. Within each segment, a model is estimated and spectral functions are calculated. In order to get smoother results, the adjacent windows may overlap. The windows should be short enough to treat the data within each of them as stationary.

Although parametric spectral estimators perform well for shorter data epochs, the statistical significance of the estimates decreases with a shortening of the window size. When we have multiple recordings of the repeated experiment at our disposal, we can incorporate the information from all the repetitions to improve the quality of the fit. Within every window, we estimate signal correlation matrices (it is a part of the model-fitting procedure); then we average these correlation matrices and find from them the model coefficients. This procedure allows for effectively increasing the number of data points used for model fitting. We assume that every repetition of our experiment is a realization of the same stochastic process. Thus, the correlation structure of the signal should be the same in every realization, giving in result the same correlation matrix estimate. This technique based on ensemble averaging allows the use of very short data segments. The window length which may be applied depends on the number of realizations, since the number of data points cannot be too small. The rule of thumb is that the number of estimated parameters should be at least several ($\sim$10) times smaller than the number of data points. The details of the procedure can be found in Ding et al. (2000) and Kamiński et al. (2001).

The DTF results obtained by the sliding window technique are termed in the literature the short-time DTF (SDTF, STDTF). That function allows for identification of the dynamically changing patterns of transmissions. By integrating the SDTF values in specific frequency ranges we may, for instance, examine the propagation of a given rhythm in the form of an animation.

2.11   EXAMPLES OF APPLICATIONS

2.11.1   MVAR Applications for Spike Train Data

Point-process data typically consist of long periods of slowly (or not at all) changing base value and quick short events such as spikes. Such signals do not seem suitable for AR model analysis. However, after some preprocessing, they may be adapted to the linear modeling formalism as was shown in Kocsis and Kamiński (2006). The paper describes investigations of the origin of the theta rhythm in the hippocampus of a rat. The hippocampus and septum are structures from which the activity was observed. Both structures are not only reciprocally anatomically connected, but they are also connected to neurons in the supramamillary nucleus (SUM), which is suspected to

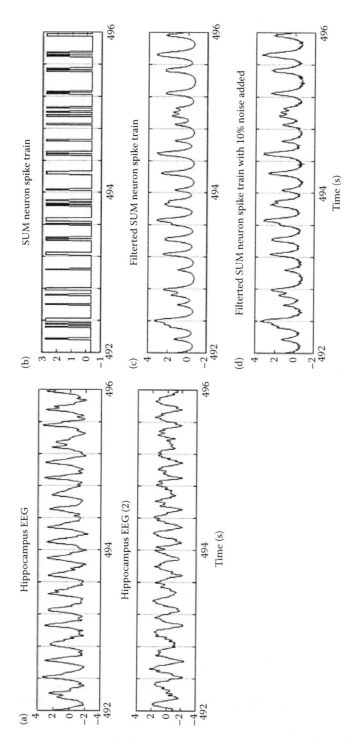

FIGURE 2.5 Preprocessing of signals for DTF. (a) Local field potentials recorded from the hippocampus using two electrodes with a vertical tip separation of 1.5 mm. (b) Standardized spike train from the SUM. (c) Low-pass filtered spike train. (d) Low-pass filtered spike train with 10% of noise added. (Reprinted from Kocsis, B. and M. Kamiński. 2006. *Hippocampus 16*: 531–540.)

be the driving structure for the theta rhythm. Moreover, during a sensory stimulation, a change appears in the theta rhythm.

To estimate the source of activity, recordings from several neurons of the SUM and LFP signals from the hippocampus were made. While hippocampal activity has a form of EEG-like signal, the SUM neurons' recordings were in the form of spike trains.

The first step of the adaptation of the SUM signals to AR modeling was filtering them by a lowpass filter (forward and backward to avoid phase shifts). Then, a white noise of amplitude of 10% of the signal was added as a necessary stochastic component. The resulting time series are presented in Figure 2.5.

The important result of the analysis is shown in Figure 2.6. Two signals, the hippocampus LFP and a SUM spike train, were analyzed using the short-time DTF method. In the first half of the signal, a sensory stimulation was performed and during that period, a change in the signal character was visible.

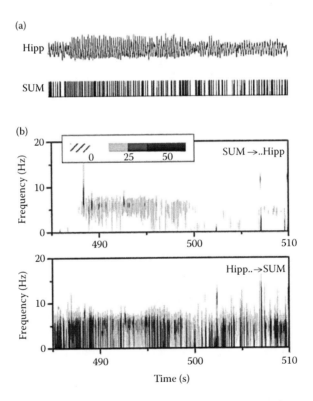

FIGURE 2.6 Activation of the theta drive directed from SUM to the hippocampus during sensory stimulation. (a) Traces of the rhythmic hippocampal EEG and SUM spike train. Note the increase in theta frequency and amplitude during the tail pinch (488–500 s). (b) Temporal dynamic of the SUM to the hippocampus flow estimated by DTF. (c) Temporal dynamics of the hippocampus to the SUM flow. (Reprinted from Kocsis, B. and M. Kamiński. 2006. *Hippocampus 16*: 531–540.)

The analysis shows that while transmission from the hippocampus to SUM can be observed in the whole epoch, a transmission from SUM to the hippocampus appears only during the sensory stimulation, at a slightly higher frequency (close to 5 Hz). This is in agreement with the expected change in the theta activity also observed in the power spectra.

The functional connectivity analysis done by SDTF showed that in these two structures, anatomically connected in both directions, the spontaneous and sensory-induced theta rhythms have different origins.

2.11.2   MVAR/SDTF APPLICATIONS IN MOTOR AND COGNITIVE EXPERIMENTS

One of the first applications of SDTF concerned the determination of the dynamic propagation during the performance of finger movement and its imagination (Ginter Jr et al., 2001a; Kuś et al., 2006). The results were in agreement with the known phenomena of event-related synchronization (ERS) and desynchronization (ERD). It was reported in papers concerning ERD/ERS (Pfurtscheller and Arnibar, 1979) that during the movement a decrease in activity in the alpha and beta bands (ERD) in the primary motor area corresponding to the given part of the body occurs and later an increase of beta activity called beta rebound follows. In the gamma band, a brief increase during the movement was reported (Pfurtscheller and Arnibar, 1979). These findings corresponded very well with the results obtained by means of SDTF. Typically, during the movement, a decrease of propagation in the alpha and beta bands was observed from the electrodes overlying the finger motor area (Ginter Jr et al., 2001a,b). Later, a fast increase of EEG propagation in the beta band from the fronto-central electrodes was noticed. The short burst of gamma propagation from the electrodes placed directly over the finger motor area was accompanying the finger movement. In case of movement imagination, this propagation started later and a cross-talk between different locations overlying the motor area and supplementary motor area was observed (Kuś et al., 2006). This kind of transmissions supports the neurophysiological hypotheses concerning the interactions of brain structures during simple and complex motor tasks. The animation showing the dynamic propagation during finger movement is available at: http://brain.fuw.edu.pl/~kjbli/DTF_MOV.html.

Other applications of SDTF concerned evaluation of transmission during cognitive experiments. The EEG propagation during the Continuous Attention Test (CAT) was investigated (Kuś et al., 2008; Blinowska et al., 2010). The CAT test relies on identification of the two identical consecutive geometrical patterns, which has to be signalized by pressing the switch (condition called target). When the consecutive pictures are different, no action is required (nontarget). During both conditions, the increased EEG transmission was found in the pre-frontal and frontal areas. In case of a nontarget, the propagation from rIFC (F8)—the right inferior cortex (in 6 subjects) or from the preSMA—pre-supplementary motor area (in 3 subjects) to the motor cortex (C3) was observed. PreSMA and rIFC are the structures presumably responsible for no-go action—withdrawal from the motor action (Figure 2.7). Our experiment confirmed the active inhibition exerted by the above structures on the motor cortex.

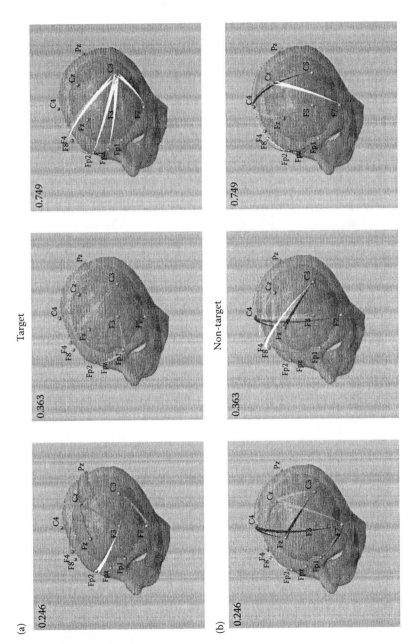

FIGURE 2.7 **(See color insert.)** Snapshots from the movie presenting significant changes in transmissions in one subject, for target (a) and nontarget (b). Intensity of flow changes for increase: from pale yellow to red, for decrease: from light to dark blue. In the right upper corner, the time after cue presentation (in seconds). (Reprinted from Blinowska et al. 2010. *Brain Topogr.* 23: 205–213.)

For the target condition, the activity burst was visible from the electrode overlying the motor cortex, which corroborates the previous experiment on the finger movement. The time-varying transmissions evaluated by means of SDTF during the CAT test are available at URL http://brain.fuw.edu.pl/~kjbli/CAT_MOV.html.

2.11.3   The SdDTF Function

In case of the signals measured by means of electrodes placed directly on/in the cortex, it is important to estimate the direct connections between the brain structures. In Korzeniewska et al. (2008), the time-varying event-related connectivity was evaluated for EEG measured by intracranial electrodes for the task involving the word repetition. In order to estimate dynamic direct propagation, an estimator, which was a time-varying version of dDTF, was introduced. It was called the short-time direct directed transfer function—SdDTF. This measure combined the benefits of directionality, directedness, and short-time windowing. It was checked by means of extensive simulations that SdDTF properly estimated the directions, spectral content, intensity, and direct causal interactions between signals and their time evolution.

The aim of the experiment was to evaluate event-related changes in ECoG transmission during the word repetition task. New statistical methodology called event-related causality (ERC) was proposed in Korzeniewska et al. (2008) to assess the significance of the propagation changes in respect to the reference epoch. The results revealed the patterns of communication between the auditory and mouth-tongue motor areas, which changed in consecutive phases of the experiment: listening, response preparation, verbal response.

The latest publications indicate that MVAR formalism can be used in investigations of causal relations between brain structures based on fMRI images as well (Harrison et al., 2003; Sato et al., 2009).

2.12   CONCLUDING REMARKS

The autoregressive modeling and causal connectivity estimation by means of DTF/PDC has been successfully used in many areas of signal analysis. Even point-process data (such as spike trains) or fMRI data may be approached by that methodology. Different measures based on DTF include dDTF, SDTF, and SdDTF. The last two estimators allow one to find time-varying transmission and bring the unique opportunity to follow the dynamical evolution of brain information processing. Some examples of DTF application include: propagation of EEG activity during different stages of sleep and wakefulness (Kamiński et al., 1997), localization of epileptogenic focus (Franaszczuk et al., 1994; Franaszczuk and Bergey, 1998), transmissions between brain structures of the behaving animal (Korzeniewska et al., 1997, 2003), investigation of epileptogenesis (Medvedev and Willoughby, 1999), estimation of functional connectivity for different paradigms (Astolfi et al., 2005; Babiloni et al., 2005), or information transfer during the transitive reasoning task (Brzezicka et al., 2011). It was demonstrated in multiple experiments that results based on the application of directed transfer function are compatible with physiological knowledge and, at the same time, they bring new information on the dynamic interaction between the brain structures.

The success of the DTF function in solving various problems is connected with its favorable properties. Typically, DTF is very robust in respect to noise and insensitive to volume conduction. The robustness to noise is shared by the estimators based on the AR model, which extracts information from the noisy background. It was demonstrated by means of simulations that the propagation is correctly identified by DTF even if the signal is embedded in noise with an amplitude three times as big as the signal itself (Kamiński and Blinowska, 1991). Another important feature of the multivariate causality measures defined in the framework of MVAR is the fact that they are not sensitive to volume conduction. DTF and PDC are based on detecting the phase difference between signals. Volume conduction is a zero phase propagation; therefore, it does not influence these measures. Similarly, the artifacts, which have the same phase in all channels, do not influence DTF.

Herein, we tried to sketch the evolution of the DTF, namely, to trace its origin and then to follow its development. Now, after 22 years since the first formulation of the DTF function, its validity and usefulness in the biomedical data analysis has been proven by many experiments. Various modifications of the original definition were proposed extending the applicability of the original function to new areas of research.

REFERENCES

Achermann, P., R. Hartmann, A. Gunzinger et al. 1984. All night sleep and artificial stochastic control signals have similar correlation dimension. *Electroenceph. clin. Neurophysiol. 90*: 34–387.

Akaike, H. 1968. On the use of a linear model for the identification of feedback systems. *Ann. Inst. Stat. Math. 20*: 425–440.

Akaike, H. 1974. A new look at statistical model identification. *IEEE Trans. Autom. Control 19*: 716–723.

Arnold, M., W. H. R. Miltner, R. Bauer et al. 1998. Multivariate autoregressive modeling of nonstationary time series by means of Kalman filtering. *IEEE Trans. Biomed. Eng. 45*: 553–562.

Astolfi, L., F. Cincotti, D. Mattia et al. 2005. Assessing cortical functional connectivity by linear inverse estimation and directed transfer function: Simulations and application to real data. *Clin. Neurophysiol. 116*: 920–932.

Babiloni, F., F. Cincotti, C. Babiloni et al. 2005. Estimation of the cortical functional connectivity with the multimodal integration of high-resolution EEG and fMRI data by directed transfer function. *NeuroImage 24*: 118–131.

Baccalá, L. A. and K. Sameshima. 2001. Partial directed coherence: A new concept in neural structure determination. *Biol. Cybern. 84*: 463–474.

Baccalá, L. A., D. Y. Takahashi, and K. Sameshima. 2007. Generalized partial directed coherence. In *15th International Conference on Digital Signal Processing*, pp. 163–166.

Benveniste, A., M. Metivier, and P. Priouret. 1990. *Adaptive Algorithms and Stochastic Approximations*. Heidelberg: Springer-Verlag.

Blinowska, K. J. and P. J. Franaszczuk. 1989. A model of the generation of electrocortical rhythms. *In Springer Series in Brain Dynamics 2*. Berlin, Heidelberg: Springer-Verlag.

Blinowska, K. J., R. Kuś, and M. Kamiński. 2004. Granger causality and information flow in multivariate processes. *Phys. Rev. E 70*: 050902.

Blinowska, K. J., R. Kuś, M. Kamiński et al. 2010. Transmission of brain activity during cognitive task. *Brain Topogr. 23*: 205–213.

Blinowska, K. J. and M. Malinowski. 1991. Non-linear and linear forecasting of the EEG time series. *Biol. Cybern. 66*: 159–165.

Box, G., G. M. Jenkins and G. Reinsel. 2008. *Time Series Analysis: Forecasting and Control.* Hoboken: Wiley.

Brovelli, A., M. Ding, A. Ledberg et al. 2004. Beta oscillations in a large-scale sensorimotor cortical network: Directional influences revealed by Granger causality. *Proc. Natl. Acad. Sci. USA 101*: 9849–9854.

Brzezicka, A., M. Kamiński, J. Kamiński et al. 2011. Information transfer during a transitive reasoning task. *Brain Topogr. 24*: 1–8.

Caines, P. E. and C. W. Chan. 1975. Feedback between stationary stochastic processes. *IEEE Trans. Autom. Control 20*: 498–508.

David, O., D. Cosmelli, and K. J. Friston. 2004. Evaluation of different measures of functional connectivity using a neural mass model. *NeuroImage 21*: 659–673.

Ding, M., S. L. Bressler, W. Yang et al. 2000. Short-window spectral analysis of cortical event related potentials by adaptive multivariate autoregressive modeling: Data preprocessing, model validation, and variability assessment. *Biol. Cybern. 83*: 35–45.

Efron, B. and R. J. Tibshirani. 1993. *An Introduction to the Bootstrap.* Boca Raton, FL: Chapman & Hall.

Franaszczuk, P. J. and K. J. Blinowska. 1985. Linear model of brain electrical activity—EEG as a superposition of damped oscillatory modes. *Biol. Cybern. 53*: 19–25.

Franaszczuk, P. J., K. J. Blinowska, and M. Kowalczyk. 1985. The application of parametric multichannel spectral estimates in the study of electrical brain activity. *Biol. Cybern. 51*: 239–247.

Franaszczuk, P. J., G. K. Bergey, and M. Kamiński. 1994. Analysis of mesial temporal seizure onset and propagation using the directed transfer function method. *Electroenceph. clin. Neurophysiol. 91*: 413–427.

Franaszczuk, P. J. and G. K. Bergey. 1998. Application of the directed transfer function method to mesial and lateral onset temporal lobe seizures. *Brain Topogr. 11*: 13–21.

Gath, I., C. Feuerstein, D. T. Pham et al. 1992. On the tracking of rapid dynamic changes in seizure EEG. *IEEE Trans. Biomed. Eng. 39*: 952–958.

Gersch, W. 1987. Non-stationary multichannel time series analysis. In *Handbook of Electroencephalography and Clinical Neurophysiology; Vol. 1: Methods of Analysis of Brain Electrical and Magnetic Signal*, pp. 261–296. Amsterdam: Elsevier.

Gevers, M. R. and B. D. O. Anderson. 1981. Representations of jointly stationary stochastic feedback processes. *Int. J. Control 33*: 777–809.

Geweke, J. 1982. Measurement of linear dependence and feedback between multiple time series. *J. Am. Stat. Assoc. 77*: 304–324.

Geweke, J. 1984. Measures of conditional linear dependence and feedback between time series. *J. Am. Stat. Assoc. 79*: 907–915.

Ginter Jr, J., K. J. Blinowska, M. Kamiński et al. 2001a. Phase and amplitude analysis in time frequency space—Application to voluntary finger movement. *J. Neurosci. Methods 110*: 113–124.

Ginter Jr, J., M. Kamiński, and K. J. Blinowska. 2001b. Determination of EEG activity propagation during voluntary finger movement. *Techn. Health Care 9*: 169–170.

Granger, C. W. J. 1969. Investigating causal relations in by econometric models and cross-spectral methods. *Econometrica 37*: 424–438.

Hamilton, J. D. 1994. *Time Series Analysis*. Princeton: Cambridge University Press.

Harrison, L., W. D. Penny, and K. J. Friston 2003. Multivariate autoregressive modeling of fMRI time series. *NeuroImage 19*: 1477–1491.

Harrison, M. A. F., I. Osorio, M. G. Frei et al. 2005. Correlation dimension and integral do not predict epileptic seizures. *Chaos 15*, 033106.

Hesse, W., E. Möller, M. Arnold et al. 2003. The use of time-variant EEG Granger causality for inspecting directed interdependencies of neural assemblies. *J. Neurosci. Methods 124*: 27–44.

Jenkins, G.M. and D. G. Watts. 1968. *Spectral Analysis and Its Applications*. San Francisco: Holden-Day.

Kamiński, M. and K. J. Blinowska. 1991. A new method of the description of the information flow in brain structures. *Biol. Cybern. 65*: 203–210.

Kamiński, M., K. J. Blinowska, and W. Szelenberger. 1997. Topographic analysis of coherence and propagation of EEG activity during sleep and wakefulness. *Electroenceph. Clin. Neurophysiol. 102*: 216–227.

Kamiński, M., M. Ding, W. Truccolo et al. 2001. Evaluating causal relations in neural systems: Granger causality, directed transfer function and statistical assessment of significance. *Biol. Cybern. 85*: 145–157.

Kamiński, M. 2005. Determination of transmission patterns in multichannel data. *Philos. Trans. R. Soc. B 360*: 947–952.

Kamiński, M. and H. Liang. 2005. Causal influence: Advances in neurosignal analysis. *Crit. Rev. Biomed. Eng. 33*: 347–430.

Kay, S. M. 1988. *Modern Spectral Estimation*. Englewood Cliffs, NJ: Prentice-Hall.

Kocsis, B. and M. Kamiński. 2006. Dynamic changes in the direction of the theta rhythmic drive between supramammillary nucleus and the septohippocampal system. *Hippocampus 16*: 531–540.

Korzeniewska, A., C. Crainiceanu, R. Kuś et al. 2008. Dynamics of event related causality (erc) in brain electrical activity. *Hum. Brain Mapp. 29*: 1170–1192.

Korzeniewska, A., S. Kasicki, M. Kamiński et al. 1997. Information flow between hippocampus and related structures during various types of rat's behavior. *J. Neurosci. Methods 73*: 49–60.

Korzeniewska, A., M. Mańczak, M. Kamiński et al. 2003. Determination of information flow direction between brain structures by a modified directed transfer function method (dDTF). *J. Neurosci. Methods 125*: 195–207.

Kuś, R., M. Kamiński, and K. J. Blinowska. 2004. Determination of EEG activity propagation: Pair-wise versus multichannel estimate. *IEEE Trans. Biomed. Eng. 51*: 1501–1510.

Kuś, R., J. S. Ginter and K. J. Blinowska. 2006. Propagation of EEG activity during finger movement and its imagination. *Acta Neurobiol. Exp. 66*: 195–206.

Kuś, R., K. J. Blinowska, M. Kamiński et al. 2008. A transmission of information during continuous attention test. *Acta Neurobiol. Exp. 68*: 103–112.

Lai, Y. C., M. A. Harrison, M. G. Frei et al. 2003. Inability of Lyapunov exponents to predict epileptic seizures. *Phys. Rev. Lett. 91*, 068102.

Lütkepohl, H. 1993. *Introduction to Multiple Time Series Analysis*. New York: Springer.

Marple Jr, S. L. 1987. *Digital Spectral Analysis with Applications*. New Jersey: Prentice-Hall.

McIntosh, A. R. and F. Gonzalez-Lima. 1994. Structural equation modeling and its application to network analysis of functional brain imaging. *Hum. Brain Mapp. 2*: 2–22.

McQuarrie, A. D. R. and C. L. Tsai. 1998. *Regression and Time Series Model Selection*. Singapore: World Scientific Pub. Co. Inc.

McSharry, P. E., L. A. Smith and L. Tarassenko. 2003. Prediction of epileptic seizures: Are nonlinear methods relevant? *Nat. Med. 9*: 241–242.

Medvedev, J. and O. Willoughby. 1999. Autoregressive modeling of the EEG in systemic kainic acid induced epileptogenesis. *Internat. J. Neurosci. 97*: 149–167.

Netoff, T., L. Caroll, L. M. Pecora et al. 2006. Detecting coupling in the presence of noise and nonlinearity. In *Handbook of Time Series Analysis*, eds. B. Schelter, M. Winterhalder and J. Timmer. Weinheim: Wiley-VCH Verlag.

Pfurtscheller, G. and A. Arnibar. 1979. Evaluation of even-related desynchronization (ERD) preceding and following voluntary self paced movements. *Electroenceph. Clin. Neurophysiol. 46*: 138–146.

Pijn, J. P., L. Van Neerven, A. Noest et al. 1991. Chaos or noise in EEG signals: Dependence on state and brain site. *Electroenceph. clin. Neurophysiol. 79*: 371–381.

Pijn, J. P., D. N. Velis, M. J. van der Heyden et al. 1997. Nonlinear dynamics of epileptic seizures on basis of intracranial EEG recordings. *Brain Topogr. 9*: 249–270.

Priestley, M. B. 1983. *Spectral Analysis and Time Series*. London: Academic Press.

Saito, Y. and H. Harashima. 1981. Tracking of information within multichannel record: Causal analysis in EEG. In *Recent Advances in EEG and EMG Data Processing*, pp. 133–146. Amsterdam: Elsevier.

Sato, J. R., D. Y. Takahashi, S. M. Arcuri et al. 2009. Frequency domain connectivity identification: An application of partial directed coherence in fMRI. *Hum. Brain Mapp. 30*: 452–461.

Schelter, B., M. Winterhalder and J. Timmer (Eds.). 2006. *Handbook of Time Series Analysis*. Weinheim: Wiley-VCH Verlag.

Schneider, T. and A. Neumaier. 2001. Estimation of parameters and eigenmodes of multivariate autoregressive models. *ACM Trans. Math. Soft. 27*: 27–65.

Simon, D. 2006. *Optimal State Estimation: Kalman, H Infinity, and Nonlinear Approaches*. Hoboken: Wiley-Interscience.

Stam, C., J. P. Pijn, P. Suffczyński et al. 1999. Dynamics of the human alpha rhythm: Evidence for non-linearity? *Clin. Neurophys. 110*: 1801–1813.

Ulrych, T. J. and T. N. Bishop. 1975. Maximum entropy spectral analysis and autoregressive decomposition. *Rev. Geophys. Space Phys. 13*: 183–200.

Wang, G., M. Takigawa and T. Matsushita. 1992. Correlation of alpha activity between the frontal and occipital cortex. *Jap. J. Physiol. 42*: 1–13.

Wendling, F., K. Ansari-Asl, F. Bartolomei et al. 2009. From EEG signals to brain connectivity: A model-based evaluation of interdependence measures. *J. Neurosci. Methods 183*: 9–18.

Wiener, N. 1956. The theory of prediction. In *Modern Mathematics for Engineers*. New York: McGraw-Hill.

Winterhalder, M., B. Schelter, W. Hesse et al. 2005. Comparison of linear signal processing techniques to infer directed interactions in multivariate neural systems. *Signal Process. 85*: 2137–2160.

3 An Overview of Vector Autoregressive Models

Pedro A. Morettin

CONTENTS

3.1 INTRODUCTION

In this chapter, we describe models for a vector time series $\mathbf{X}_t$, with n components $X_{1t}, X_{2t}, \ldots, X_{nt}$, observed in times $t = 0, \pm 1, \pm 2, \ldots$. Besides analyzing individual time series X_{it}, for which the autocorrelation contained in each series is important, we will be interested in dynamic relationships between the component series. We use the notation $\mathbf{X}_t = (X_{1t}, X_{2t}, \ldots, X_{nt})'$, $t \in \mathbb{Z}$ and X_{it} or $X_{i,t}$, for the i-th component, $i = 1, \ldots, n$, where $\mathbb{Z}$ is the set of all integers.

EXAMPLE 3.1

We can assume the vector $\mathbf{X}_t$ with components $X_{it}, i = 1, 2, 3, 4$, representing four channels $C_1, \ldots, C_4$ of an EEG epoch, taken from a patient undergoing an epileptic seizure. See Figure 3.1.

The *mean vector* of $\mathbf{X}_t$ will be denoted by

$$\boldsymbol{\mu}_t = E(\mathbf{X}_t) = (\mu_{1t}, \mu_{2t}, \ldots, \mu_{nt})' \tag{3.1}$$

and depends, in general, on t.

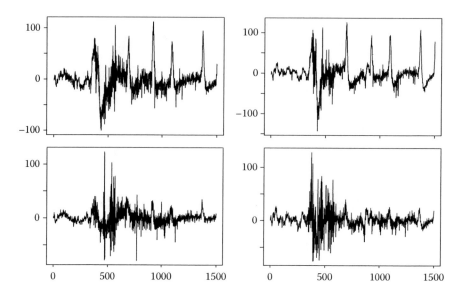

FIGURE 3.1   Four EEG channels.

The *covariance matrix* of $\mathbf{X}_t$ is defined by

$$\mathbf{\Gamma}(t + \tau, t) = E\{(\mathbf{X}_{t+\tau} - \boldsymbol{\mu}_{t+\tau})(\mathbf{X}_t - \boldsymbol{\mu}_t)'\}, \tag{3.2}$$

an $n \times n$ matrix that, in general, also depends on t.

The quantities (3.1) and (3.2) describe the second-order properties of the series $X_{1t}, \ldots, X_{nt}$. If these series have a multivariate normal distribution, these properties will be completely specified by means and covariances. Notice that Equation 3.2 gives the autocovariances of the individual series as well as the cross-covariances between pairs of different series.

If we denote by $\gamma_{ij}(t + \tau, t)$, $i, j = 1, \ldots, n$, the components of the matrix $\mathbf{\Gamma}(t + \tau, t)$, then

$$\gamma_{ij}(t + \tau, t) = \text{Cov}\{X_{i,t+\tau}, X_{j,t}\}$$
$$= E\{(X_{i,t+\tau} - \mu_{i,t+\tau})(X_{j,t} - \mu_{j,t})\}, \tag{3.3}$$

$i, j = 1, \ldots, n$, is the covariance between the series $X_{i,t+\tau}$ and $X_{j,t}$.

EXAMPLE 3.2

In Example 3.1, with $\mathbf{X}_t = (X_{1t}, \ldots, X_{4t})'$, $\boldsymbol{\mu}_t = (\mu_{1t}, \ldots, \mu_{4t})'$ is the mean vector and the matrix (3.2) becomes

$$\mathbf{\Gamma}(t + \tau, t) = \begin{bmatrix} \gamma_{11}(t + \tau, t) & \gamma_{12}(t + \tau, t) & \cdots & \gamma_{14}(t + \tau, t) \\ \gamma_{21}(t + \tau, t) & \gamma_{22}(t + \tau, t) & \cdots & \gamma_{24}(t + \tau, t) \\ \cdots & \cdots & & \cdots \\ \gamma_{41}(t + \tau, t) & \gamma_{42}(t + \tau, t) & \cdots & \gamma_{44}(t + \tau, t) \end{bmatrix}.$$

In the main diagonal, we have the autocovariances of the individual series, computed at times $t + \tau$ and t, while outside the main diagonal we have the *cross-covariance* between the series $X_{i,t+\tau}$ and $X_{j,t}$, $i \neq j$.

An important case is when the mean vector and the covariance matrix do not depend on t. We obtain (weakly) stationary series.

3.2 STATIONARY SERIES

We restrict ourselves in this chapter to weak (or second order) stationarity. We say that the n-variate series $\mathbf{X}_t$ is *stationary* if the mean $\boldsymbol{\mu}_t$ and the covariance matrix $\boldsymbol{\Gamma}(t + \tau, t), t, \tau \in \mathbb{Z}$ do not depend on t. In this situation, we have

$$\boldsymbol{\mu} = E(\mathbf{X}_t) = (\mu_1, \ldots, \mu_n)', \tag{3.4}$$

and

$$\boldsymbol{\Gamma}(\tau) = E\{(\mathbf{X}_{t+\tau} - \boldsymbol{\mu})(\mathbf{X}_t - \boldsymbol{\mu})'\} = [\gamma_{ij}(\tau)]_{i,j=1}^n, \tag{3.5}$$

$\tau \in \mathbb{Z}$. Here, $\gamma_{ii}(\tau)$ denotes the autocovariance function of the stationary series X_{it} and $\gamma_{ij}(\tau)$ denotes the cross-covariance between X_{it} and $X_{j,t+\tau}$. Notice that in general $\gamma_{ij}(\tau) \neq \gamma_{ji}(\tau)$.

If we take $\tau = 0$ in Equation 3.5, we obtain

$$\boldsymbol{\Gamma}(0) = E\{(\mathbf{X}_t - \boldsymbol{\mu})(\mathbf{X}_t - \boldsymbol{\mu})'\}, \tag{3.6}$$

called the *contemporary covariance matrix*. Notice that $\gamma_{ii}(0) = \text{Var}(X_{it})$, $\gamma_{ij}(0) = \text{Cov}\{X_{it}, X_{jt}\}$.

It follows that the *contemporary correlation coefficient* between X_{it} and X_{jt} is given by

$$\rho_{ij}(0) = \frac{\gamma_{ij}(0)}{[\gamma_{ii}(0)\gamma_{jj}(0)]^{1/2}}. \tag{3.7}$$

Obviously, $\rho_{ij}(0) = \rho_{ji}(0)$, $\rho_{ii}(0) = 1$ and $-1 \leq \rho_{ij}(0) \leq 1$, for all $i, j = 1, \ldots, n$, from which we deduce that $\boldsymbol{\rho}(0) = [\rho_{ij}(0)]_{i,j=1}^n$ is a symmetric matrix, with all elements in the diagonal matrix equal to 1.

The *correlation matrix* of lag τ is defined by

$$\boldsymbol{\rho}(\tau) = \mathbf{D}^{-1}\boldsymbol{\Gamma}(\tau)\mathbf{D}^{-1}, \tag{3.8}$$

where $\mathbf{D} = \text{diag}\{\sqrt{\gamma_{11}(0)}, \ldots, \sqrt{\gamma_{nn}(0)}\}$. Or, denoting by $\boldsymbol{\rho}(\tau) = [\rho_{ij}(\tau)]_{i,j=1}^n$, we have

$$\rho_{ij}(\tau) = \frac{\gamma_{ij}(\tau)}{[\gamma_{ii}(0)\gamma_{jj}(0)]^{1/2}}, \tag{3.9}$$

which is the correlation coefficient between $X_{i,t+\tau}$ and $X_{j,t}$.

When $\tau > 0$, this coefficient measures the linear dependence of X_{it} on X_{jt}, which occurred before time $t + \tau$. Therefore, if $\rho_{ij}(\tau) \neq 0$, $\tau > 0$, we say that X_{jt} is *leading* X_{it} at lag τ. Analogously, $\rho_{ji}(\tau)$ measures the linear dependence of X_{jt} on X_{it}, $\tau > 0$.

The fact that $\rho_{ij}(\tau) \neq \rho_{ji}(\tau)$, for all i, j, comes from the fact that these two correlation coefficients measure different linear relationships between X_{it} and X_{jt}. The matrices $\mathbf{\Gamma}(\tau)$ and $\mathbf{\rho}(\tau)$ are not, in general, symmetric. In fact, the following result.

Proposition 3.1   The following properties hold for the covariance matrix:

   i. $\mathbf{\Gamma}(\tau) = \mathbf{\Gamma}'(-\tau)$
   ii. $|\gamma_{ij}(\tau)| \leq [\gamma_{ii}(0)\gamma_{jj}(0)]^{1/2}$, $i, j = 1, \ldots, n$
   iii. $\gamma_{ii}(\tau)$ is an autocovariance function, for all i
   iv. $\sum_{j,k=1}^{m} \mathbf{a}_j' \mathbf{\Gamma}(j - k)\mathbf{a}_k \geq 0$, for all m and vectors $\mathbf{a}_1, \ldots, \mathbf{a}_m$ of $\mathbb{R}^n$.

The proofs are immediate. Notice that from (i) we obtain that $\gamma_{ij}(\tau) = \gamma_{ji}(-\tau)$. The matrix $\mathbf{\rho}(\tau)$ has similar properties, with $\rho_{ii}(0) = 1$. We also remark that $\rho_{ij}(0)$ need not be equal to one and it is also possible that $|\gamma_{ij}(\tau)| > |\gamma_{ij}(0)|$, if $i \neq j$.

We say that the $n \times n$ series $\{\mathbf{a}_t, t \in Z\}$ is a *multivariate white noise*, with mean $\mathbf{0}$ and covariance matrix $\mathbf{\Sigma}$, if $\mathbf{a}_t$ is stationary with mean $\mathbf{0}$ and its covariance matrix is given by

$$\mathbf{\Gamma}(\tau) = \begin{cases} \mathbf{\Sigma}, & \text{if } \tau = 0, \\ \mathbf{0}, & \text{if } \tau \neq 0. \end{cases} \tag{3.10}$$

We will use the notation $\mathbf{a}_t \sim \text{WN}(\mathbf{0}, \mathbf{\Sigma})$. If the vectors $\mathbf{a}_t$ are independent and identically distributed, we write $\mathbf{a}_t \sim \text{IID}(\mathbf{0}, \mathbf{\Sigma})$.

A process $\mathbf{X}_t$ is said to be *linear* if

$$\mathbf{X}_t = \sum_{j=0}^{\infty} \Psi_j \mathbf{a}_{t-j}, \tag{3.11}$$

where $\mathbf{a}_t$ is a multivariate white noise and Ψ_j is a sequence of matrices whose components are absolutely summable. It follows that $E(\mathbf{X}_t) = \mathbf{0}$ and the covariance matrix of $\mathbf{X}_t$ is given by

$$\mathbf{\Gamma}(\tau) = \sum_{j=0}^{\infty} \Psi_{j+\tau} \mathbf{\Sigma} \Psi_j', \quad \tau \in \mathbb{Z}. \tag{3.12}$$

3.3   ESTIMATION OF MEANS AND COVARIANCES

Assuming that we have observations $\{\mathbf{X}_t, t = 1, \ldots, T\}$ from the stationary process $\{\mathbf{X}_t, t \in \mathbb{Z}\}$, the mean $\mathbf{\mu}$ is estimated by the vector of sample means

$$\overline{\mathbf{X}} = \frac{\sum_{t=1}^{T} \mathbf{X}_t}{T}. \tag{3.13}$$

It follows that the mean μ_j of X_{jt} is estimated by $\sum_{t=1}^{T} X_{jt}/T$.

It can be proved, under several conditions on the process $\mathbf{X}_t$, that (Brockwell and Davis, 1991):

   i. $E(\overline{\mathbf{X}} - \mathbf{\mu})'(\overline{\mathbf{X}} - \mathbf{\mu}) \to 0$, if $\gamma_{ii}(\tau) \to 0$, $i = 1, \ldots, n$.

ii. $TE(\overline{\mathbf{X}} - \boldsymbol{\mu})'(\overline{\mathbf{X}} - \boldsymbol{\mu}) \to \sum_{i=1}^{n} \sum_{\tau} \gamma_{ii}(\tau)$, if $\sum_{\tau} |\gamma_{ii}(\tau)| < \infty$, $i = 1, \ldots, n$.

iii. The vector $\overline{\mathbf{X}}$ is asymptotically normally distributed.

In order to estimate $\Gamma(\tau)$, we use

$$\hat{\Gamma}(\tau) = \begin{cases} \dfrac{1}{T} \displaystyle\sum_{t=1}^{T-\tau} (\mathbf{X}_{t+\tau} - \overline{\mathbf{X}})(\mathbf{X}_t - \overline{\mathbf{X}})', & 0 \le \tau \le T-1 \\[2em] \dfrac{1}{T} \displaystyle\sum_{t=-\tau+1}^{T} (\mathbf{X}_{t+\tau} - \overline{\mathbf{X}})(\mathbf{X}_t - \overline{\mathbf{X}})', & -T+1 \le \tau \le 0. \end{cases} \tag{3.14}$$

Consequently, the correlation matrix may be estimated by

$$\hat{\rho}(\tau) = \hat{\mathbf{D}}^{-1} \hat{\Gamma}(\tau) \hat{\mathbf{D}}^{-1}, \tag{3.15}$$

where $\hat{\mathbf{D}}$ is the $n \times n$ diagonal matrix containing the sample standard deviations of the individual series.

See Fuller (1996) for properties of the estimators $\hat{\Gamma}(\tau)$ and $\hat{\rho}(\tau)$.

For the computations of the correlation matrices of the next examples, and further computations done in this chapter, we have used the software SPlus. Other possibilities are MATLAB and the repository of packages R (Comprehensive R Archive Network, CRAN). Details in http://CRAN.R-project.org.

EXAMPLE 3.3

Suppose that X_{1t} and $X_2(t)$ represent the first $T = 1500$ observations of the EEG series C_1 and C_2 of Example 3.1. Let $\mathbf{X}_t = (X_{1t}, X_{2t})'$. Assume $\mathbf{X}_t$ is stationary. Table 3.1 shows some sample correlation matrices. A convenient way to represent these matrices is by using the symbols $+$, $-$ and $\cdot$, when a correlation value is, respectively, greater or equal to $2/\sqrt{T}$, smaller or equal to $-2/\sqrt{T}$ or lies between $-2/\sqrt{T}$ and $2/\sqrt{T}$. These simplified matrices are also shown in Table 3.1.

We see, for example, that

$$\hat{\rho}(0) = \begin{bmatrix} 1.000 & 0.792 \\ 0.792 & 1.000 \end{bmatrix},$$

while

$$\hat{\rho}(1) = \begin{bmatrix} 0.784 & 0.735 \\ 0.726 & 0.820 \end{bmatrix}.$$

Since $2/\sqrt{1\,500} = 0.052$, all correlations of lag one are statistically significant, and the simplified representation of this matrix is

$$\begin{bmatrix} + & + \\ + & + \end{bmatrix}.$$

TABLE 3.1

Sample Correlation Matrices for Example 3.2, with Simplified Notation

Lag 1	Lag 2	Lag 3	Lag 4
$\begin{bmatrix} 0.784 & 0.735 \\ 0.726 & 0.820 \end{bmatrix}$	$\begin{bmatrix} 0.776 & 0.734 \\ 0.710 & 0.828 \end{bmatrix}$	$\begin{bmatrix} 0.758 & 0.724 \\ 0.692 & 0.800 \end{bmatrix}$	$\begin{bmatrix} 0.736 & 0.712 \\ 0.645 & 0.770 \end{bmatrix}$
$\begin{bmatrix} + & + \\ + & + \end{bmatrix}$	$\begin{bmatrix} + & + \\ + & + \end{bmatrix}$	$\begin{bmatrix} + & + \\ + & + \end{bmatrix}$	$\begin{bmatrix} + & + \\ + & + \end{bmatrix}$

3.4 VECTOR AUTOREGRESSIVE MODELS

In this section, we study an important class of linear multivariate models, namely, that of vector autoregressive models, denoted briefly by VAR(p). We say that the process $\mathbf{X}_t$, of order $n \times 1$, follows a VAR(p) model if

$$\mathbf{X}_t = \mathbf{\Phi}_0 + \mathbf{\Phi}_1 \mathbf{X}_{t-1} + \cdots + \mathbf{\Phi}_p \mathbf{X}_{t-p} + \mathbf{a}_t, \tag{3.16}$$

where $\mathbf{a}_t \sim \text{WN}(\mathbf{0}, \mathbf{\Sigma})$, $\mathbf{\Phi}_0 = (\phi_{10}, \dots, \phi_{n0})'$ is an $n \times 1$ vector of constants and $\mathbf{\Phi}_k$ is an $n \times n$ matrix of constants, with elements $\phi_{ij}^{(k)}$, $i, j = 1, \dots, n$, $k = 1, \dots, p$. If $\mathbf{I}_n$ is the identity matrix of order n, model (3.16) can be written in the form

$$\mathbf{\Phi}(B)\mathbf{X}_t = \mathbf{\Phi}_0 + \mathbf{a}_t, \tag{3.17}$$

where $\mathbf{\Phi}(B) = \mathbf{I}_n - \mathbf{\Phi}_1 B - \cdots - \mathbf{\Phi}_p B^p$ is the vector autoregressive operator of order p, or, a polynomial matrix of order n in B. The generic element of $\mathbf{\Phi}(B)$ is $[\delta_{ij} - \phi_{ij}^{(1)} B - \cdots - \phi_{ij}^{(p)} B^p]$, for $i, j = 1, \dots, n$ and $\delta_{ij} = 1$, if $i = j$ and equal to zero otherwise.

We consider now, for simplicity, the VAR(1) model, namely

$$\mathbf{X}_t = \mathbf{\Phi}_0 + \mathbf{\Phi}\mathbf{X}_{t-1} + \mathbf{a}_t. \tag{3.18}$$

An important special case is when $n = 2$ and Equation 3.18 reduces to

$$X_{1t} = \phi_{10} + \phi_{11}X_{1,t-1} + \phi_{12}X_{2,t-1} + a_{1t},$$
$$X_{2t} = \phi_{20} + \phi_{21}X_{1,t-1} + \phi_{22}X_{2,t-1} + a_{2t}, \tag{3.19}$$

where we neglect the index 1 in $\mathbf{\Phi}_1$ and in $\phi_{ij}^{(1)}$. Denote the elements of $\mathbf{\Sigma}$ by σ_{ij}, $i, j = 1, 2$.

Observe that Equation 3.19 does not make explicit the concurrent dependence between X_{1t} and X_{2t}. We say that Equations 3.19 and 3.18 are in *reduced* form. It is possible to obtain the model in the *structural* form, where this relationship is made explicit. For this, a Cholesky decomposition has to be applied to the model. See Tsay (2005) for details.

Let us go back to Equation 3.19. If $\phi_{12} = 0$, the series X_{1t} does not depend on $X_{2,t-1}$ and similarly if $\phi_{21} = 0$, $X_{2,t}$ does not depend on $X_{1,t-1}$. On the other hand, if $\phi_{12} = 0$ and $\phi_{21} \neq 0$, there exists a linear relationship from X_{1t} to X_{2t}. If $\phi_{12} = \phi_{21} = 0$, we say that there is not a linear relationship between the series, or that they are *not coupled*. Finally, if $\phi_{12} \neq 0$, $\phi_{21} \neq 0$, we say that there is *feedback* between the series. Moreover, if $\sigma_{12} = 0$ in $\mathbf{\Sigma}$, there is no linear concurrent relationship between X_{1t} and X_{2t}.

The process $\mathbf{X}_t$ in Equation 3.18 is stationary if the mean is constant and $E(\mathbf{X}_{t+\tau}\mathbf{X}_t')$ is independent of t. In this case, if $\boldsymbol{\mu} = E(\mathbf{X}_t)$, we will have

$$\boldsymbol{\mu} = (\mathbf{I}_n - \mathbf{\Phi})^{-1}\mathbf{\Phi}_0.$$

It follows that the model can be written in the form

$$\mathbf{X}_t - \boldsymbol{\mu} = \mathbf{\Phi}(\mathbf{X}_{t-1} - \boldsymbol{\mu}) + \mathbf{a}_t,$$

or, if $\tilde{\mathbf{X}}_t = \mathbf{X}_t - \boldsymbol{\mu}$,

$$\tilde{\mathbf{X}}_t = \mathbf{\Phi}\tilde{\mathbf{X}}_{t-1} + \mathbf{a}_t. \tag{3.20}$$

As in the case of an AR(1) (univariate) model, we obtain from Equation 3.20

$$\tilde{\mathbf{X}}_t = \mathbf{a}_t + \mathbf{\Phi}\mathbf{a}_{t-1} + \mathbf{\Phi}^2\mathbf{a}_{t-2} + \cdots, \tag{3.21}$$

which is the infinite moving average representation (MA(∞)) of the model. It is also easy to see that $\text{Cov}(\mathbf{a}_t, \mathbf{X}_{t-1}) = \mathbf{0}$ and $\text{Cov}(\mathbf{a}_t, \mathbf{X}_t) = \mathbf{\Sigma}$.

Denote by $|\mathbf{A}|$ the determinant of the square matrix $\mathbf{A}$.

Proposition 3.2 The process $\mathbf{X}_t$ following a VAR(1) model will be stationary if all the solutions of

$$|\mathbf{I}_n - \mathbf{\Phi}z| = 0 \tag{3.22}$$

are outside the unit circle.

Since the solutions of Equation 3.22 are the inverse of the eigenvalues of $\mathbf{\Phi}$, an equivalent condition is that all the eigenvalues of $\mathbf{\Phi}$ should be in absolute value less than 1. Or $|\mathbf{I}_n - \mathbf{\Phi}z| \neq 0$, $|z| \leq 1$.

EXAMPLE 3.4

In the case of a bivariate VAR(1), Equation 3.22 becomes

$$\begin{vmatrix} 1 - \phi_{11}z & -\phi_{12}z \\ -\phi_{21}z & 1 - \phi_{22}z \end{vmatrix} = (1 - \phi_{11}z)(1 - \phi_{22}z) - \phi_{12}\phi_{21}z^2 = 0,$$

that is, we get the equation

$$1 - \text{tr}(\mathbf{\Phi})z - |\mathbf{\Phi}|z^2 = 0,$$

where $\text{tr}(\boldsymbol{\Phi}) = \phi_{11} + \phi_{22}$ indicates the trace of $\boldsymbol{\Phi}$. For example, if

$$\boldsymbol{\Phi} = \begin{bmatrix} 0.5 & 0.3 \\ -0.6 & -0.1 \end{bmatrix},$$

then $\text{tr}(\boldsymbol{\Phi}) = 0.4$, $|\boldsymbol{\Phi}| = 0.13$ and the roots have a modulus greater than 1.

EXAMPLE 3.5

Consider the VAR (1) model ($n = 2$)

$$X_{1,t} = 0.4 + 0.5X_{1,t-1} + 0.3X_{2,t-1} + a_{1,t},$$
$$X_{2,t} = -1.7 - 0.6X_{1,t-1} - 0.1X_{2,t-1} + a_{2,t},$$

with

$$\boldsymbol{\Sigma} = \begin{bmatrix} 1 & 0.5 \\ 0.5 & 1 \end{bmatrix}.$$

Here,

$$\boldsymbol{\Phi}_1 = \begin{bmatrix} 0.5 & 0.3 \\ -0.6 & -0.1 \end{bmatrix}, \quad \boldsymbol{\Phi}_0 = \begin{bmatrix} 0.4 \\ -1.7 \end{bmatrix}, \quad \boldsymbol{\mu} = \begin{bmatrix} 2.0 \\ -1.0 \end{bmatrix}.$$

Figure 3.2 shows a simulation of these two series. It is easy to check that the model is stationary.

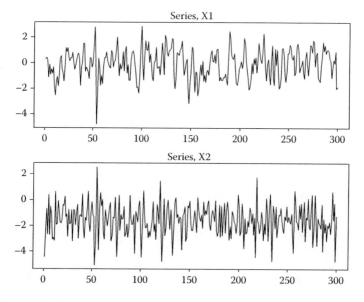

FIGURE 3.2 Simulated VAR(1) model.

Now we compute the covariance matrix of $\mathbf{X}_t$, under model (3.20). Using Equation 3.21, we have

$$\mathbf{\Gamma}(0) = \mathbf{\Sigma} + \mathbf{\Phi}\mathbf{\Sigma}\mathbf{\Phi}' + \mathbf{\Phi}^2\mathbf{\Sigma}(\mathbf{\Phi}^2)' + \cdots = \sum_{j=0}^{\infty} \mathbf{\Phi}^j\mathbf{\Sigma}(\mathbf{\Phi}^j)', \quad \mathbf{\Phi}_0^0 = \mathbf{I}_n.$$

A similar formula holds for $\mathbf{\Gamma}(\tau)$. But these formulas involve infinite sums, which are not very useful for calculations. If we post-multiply (3.20) by $\tilde{\mathbf{X}}_{t-\tau}'$ and take expectations, we get

$$E(\tilde{\mathbf{X}}_t\tilde{\mathbf{X}}_{t-\tau}') = \mathbf{\Phi}E(\tilde{\mathbf{X}}_{t-1}\tilde{\mathbf{X}}_{t-\tau}') + E(\mathbf{a}_t\tilde{\mathbf{X}}_{t-\tau}').$$

For $\tau = 0$, we obtain

$$\mathbf{\Gamma}(0) = \mathbf{\Phi}\mathbf{\Gamma}(-1) + \mathbf{\Sigma} = \mathbf{\Phi}\mathbf{\Gamma}(1)' + \mathbf{\Sigma}.$$

This means that for computing $\mathbf{\Gamma}(0)$, we need $\mathbf{\Phi}$, $\mathbf{\Sigma}$, and $\mathbf{\Gamma}(1)$. Since $E(\mathbf{a}_t\tilde{\mathbf{X}}_{t-\tau}') = 0$, for $\tau > 0$, it follows that

$$\mathbf{\Gamma}(\tau) = \mathbf{\Phi}\mathbf{\Gamma}(\tau - 1), \quad \tau > 0,$$

and after substitutions we find

$$\mathbf{\Gamma}(\tau) = \mathbf{\Phi}^\tau\mathbf{\Gamma}(0), \quad \tau > 0. \tag{3.23}$$

It follows that $\mathbf{\Gamma}(1) = \mathbf{\Phi}\mathbf{\Gamma}(0)$, hence

$$\mathbf{\Gamma}(0) = \mathbf{\Phi}\mathbf{\Gamma}(0)\mathbf{\Phi}' + \mathbf{\Sigma}.$$

Taking the vec operator in both sides, we get

$$\mathrm{vec}(\mathbf{\Gamma}(0)) = \mathrm{vec}(\mathbf{\Phi}\mathbf{\Gamma}(0)\mathbf{\Phi}') + \mathrm{vec}(\mathbf{\Sigma}),$$

and since $\mathrm{vec}(\mathbf{\Phi}\mathbf{\Gamma}(0)\mathbf{\Phi}') = (\mathbf{\Phi} \otimes \mathbf{\Phi})\mathrm{vec}\,(\mathbf{\Gamma}(0))$, we finally obtain

$$\mathrm{vec}(\mathbf{\Gamma}(0)) = (\mathbf{I}_{n^2} - \mathbf{\Phi} \otimes \mathbf{\Phi})^{-1}\mathrm{vec}(\mathbf{\Sigma}).$$

In this expression, the inverse exists since $|(\mathbf{I}_{n^2} - \mathbf{\Phi} \otimes \mathbf{\Phi})| \neq 0$, due to the fact that the eigenvalues of the product $\mathbf{\Phi} \otimes \mathbf{\Phi}$ are the same as the product of the eigenvalues of $\mathbf{\Phi}$, and hence have absolute values smaller than 1.

The previous results can be extended to VAR(p) models, $p > 1$, due to the fact that such a process can always be written as a VAR(1) model. It follows that the following proposition is valid.

Proposition 3.3   For the VAR(p) model given in Equation 3.16, the following results hold:

i. The process $\mathbf{X}_t$ is stationary if the roots of

$$|\mathbf{I}_n - \mathbf{\Phi}_1 z - \cdots - \mathbf{\Phi}_p z^p| = 0$$

are all outside the unit circle.

ii. If $\mathbf{X}_t$ is stationary,

$$\boldsymbol{\mu} = E(\mathbf{X}_t) = (\mathbf{I}_n - \mathbf{\Phi}_1 z - \cdots - \mathbf{\Phi}_p z^p)^{-1} \mathbf{\Phi}_0.$$

iii. Writing Equation 3.16 in the form

$$\tilde{\mathbf{X}}_t = \mathbf{\Phi}_1 \tilde{\mathbf{X}}_{t-1} + \cdots + \mathbf{\Phi}_p \tilde{\mathbf{X}}_{t-p} + \mathbf{a}_t,$$

with $\tilde{\mathbf{X}}_t = \mathbf{X}_t - \boldsymbol{\mu}$ and multiplying this equation by $\tilde{\mathbf{X}}'_{t-\tau}$ we obtain

$$\mathbf{\Gamma}(\tau) = \mathbf{\Phi}_1 \mathbf{\Gamma}(\tau - 1) + \cdots + \mathbf{\Phi}_p \mathbf{\Gamma}(\tau - p), \quad \tau > 0,$$

which are the Yule–Walker equations for a VAR(p) model.

Let us remark that

$$\mathbf{\Gamma}(0) = \mathbf{\Phi}_1 \mathbf{\Gamma}(-1) + \cdots + \mathbf{\Phi}_p \mathbf{\Gamma}(-p) + \mathbf{\Sigma}$$
$$= \mathbf{\Phi}_1 \mathbf{\Gamma}(1)' + \cdots + \mathbf{\Phi}_p \mathbf{\Gamma}(p)' + \mathbf{\Sigma}.$$

These equations can be used to compute $\mathbf{\Gamma}(\tau)$ recursively, for $\tau \geq p$. For $|\tau| < p$, we need to use the VAR(1) representation of a VAR(p) process.

3.5   CONSTRUCTION OF VAR MODELS

The construction of VAR models follows the same cycle of identification, estimation, and diagnosis used for other families of models.

3.5.1   IDENTIFICATION

One way to identify the order p of a VAR(p) model consists in fitting sequentially VAR of orders $1, 2, \ldots, k$ and then testing the significance of the coefficients (matrices). Consider then the models

$$\mathbf{X}_t = \mathbf{\Phi}_0{}^{(1)} + \mathbf{\Phi}_1{}^{(1)}\mathbf{X}_{t-1} + \mathbf{a}_t^{(1)},$$

$$\mathbf{X}_t = \mathbf{\Phi}_0{}^{(2)} + \mathbf{\Phi}_1^{(2)}\mathbf{X}_{t-1} + \mathbf{\Phi}_2^{(2)}\mathbf{X}_{t-2} + \mathbf{a}_t^{(2)},$$

$$\cdots \quad \cdots$$

$$\mathbf{X}_t = \mathbf{\Phi}_0^{(k)} + \mathbf{\Phi}_1^{(k)}\mathbf{X}_{t-1} + \cdots + \mathbf{\Phi}_k^{(k)}\mathbf{X}_{t-k} + \mathbf{a}_t^{(k)}. \tag{3.24}$$

The parameter of the models can be estimated by ordinary least squares (OLS), which yield consistent and efficient estimators. We test

$$H_0 : \mathbf{\Phi}_k^{(k)} = \mathbf{0},$$

$$H_1 : \mathbf{\Phi}_k^{(k)} \neq 0, k = 1, 2, \ldots. \tag{3.25}$$

The likelihood ratio test is based on the estimates of the residual covariance matrices of the fitted models. For the k-th equation take

$$\hat{\mathbf{a}}_t^{(k)} = \mathbf{X}_t - \hat{\mathbf{\Phi}}_0^{(k)} - \hat{\mathbf{\Phi}}_1^{(k)}\mathbf{X}_{t-1} - \cdots - \hat{\mathbf{\Phi}}_k^{(k)}\mathbf{X}_{t-k}.$$

The residual covariance matrix, which estimates $\mathbf{\Sigma}$, is given by

$$\hat{\mathbf{\Sigma}}_k = \frac{1}{T-k} \sum_{t=k+1}^{T} \hat{\mathbf{a}}_t^{(k)}(\hat{\mathbf{a}}_t^{(k)})', \ k \geq 0, \tag{3.26}$$

where for $k = 0$, $\hat{\mathbf{a}}_t^{(0)} = \mathbf{X}_t - \overline{\mathbf{X}}$. The likelihood ratio statistic for the test (3.25) is given by

$$\mathrm{RV}(k) = (T-k) \ln \frac{|\hat{\mathbf{\Sigma}}_{k-1}|}{|\hat{\mathbf{\Sigma}}_k|}, \tag{3.27}$$

which follows a chi-square distribution with n^2 degrees of freedom.

Another way to identify the order p is by using some information criterion, as

$$\mathrm{AIC}(k) = \ln(|\hat{\mathbf{\Sigma}}_k|) + 2kn^2/T \quad \text{(Akaike)},$$

$$\mathrm{BIC}(k) = \ln(|\hat{\mathbf{\Sigma}}_k|) + kn^2 \ln(T)/T \quad \text{(Schwarz)},$$

$$\mathrm{HQC}(k) = \ln(|\hat{\mathbf{\Sigma}}_k|) + kn^2 \ln(\ln(T))/T \quad \text{(Hannan–Quinn)}. \tag{3.28}$$

3.5.2 ESTIMATION

Having identified the value of p and assuming $\mathbf{a}_t \sim \mathcal{N}(\mathbf{0}, \mathbf{\Sigma})$, we can estimate the coefficients through maximum likelihood. In this case, least-squares (LS) estimates are equivalent to conditional maximum likelihood estimates (MLE).

For example, in the case of a VAR(1) model, the conditional MLE are obtained maximizing

$$\ell = -\frac{n(T+1)}{2} \ln(2\pi) + \frac{(T-1)}{2} \ln|\boldsymbol{\Sigma}^{-1}|$$

$$- \frac{1}{2} \sum_{t=2}^{T} (\mathbf{X}_t - \boldsymbol{\Phi}\mathbf{X}_{t-1})'\boldsymbol{\Sigma}^{-1}(\mathbf{X}_t - \boldsymbol{\Phi}\mathbf{X}_{t-1}), \tag{3.29}$$

obtaining

$$\hat{\boldsymbol{\Phi}} = \left[\sum_{t=2}^{T} \mathbf{X}_t\mathbf{X}_{t-1}' \right] \left[\sum_{t=2}^{T} \mathbf{X}_{t-1}\mathbf{X}_{t-1}' \right]^{-1}, \tag{3.30}$$

$$\hat{\boldsymbol{\Sigma}} = \frac{1}{T} \sum_{t=1}^{T} \hat{\mathbf{a}}_t(\hat{\mathbf{a}}_t)', \tag{3.31}$$

$$\hat{\mathbf{a}}_t = \mathbf{X}_t - \hat{\boldsymbol{\Phi}}\mathbf{X}_{t-1}. \tag{3.32}$$

In the general case, MLE are obtained by numerical optimization algorithms.

3.5.3 DIAGNOSTICS

To test if the model is adequate, we use the residuals to build the multivariate version of the Box–Ljung–Pierce statistic, given by

$$Q(m) = T^2 \sum_{\tau=1}^{m} \frac{1}{T-\tau} \mathrm{tr}(\hat{\boldsymbol{\Gamma}}(\tau)'\hat{\boldsymbol{\Gamma}}(0)^{-1}\hat{\boldsymbol{\Gamma}}(\tau)\hat{\boldsymbol{\Gamma}}(0)^{-1}), \tag{3.33}$$

and under H_0 : the series $\mathbf{a}_t$ is white noise, this statistic has a $\chi^2(n^2(m-p))$ (we should have $m > p$ for the number of degrees of freedom to be positive).

3.5.4 PREDICTION

Consider the VAR(1) model given by Equation 3.18 and suppose that the parameter $\boldsymbol{\Phi}$ is known. The prediction of X_t with origin T and horizon h is given by

$$\hat{\mathbf{X}}_T(h) = \boldsymbol{\Phi}\hat{\mathbf{X}}_T(h-1),$$

from which it follows that

$$\hat{\mathbf{X}}_T(h) = \boldsymbol{\Phi}^h\mathbf{X}_T, \quad h = 1, 2, \dots. \tag{3.34}$$

Since

$$\mathbf{X}_{t+h} = \boldsymbol{\Phi}\mathbf{X}_{T+h-1} + \mathbf{a}_{T+h},$$

the prediction error h steps ahead is given by

$$\mathbf{e}_T(h) = \mathbf{X}_{t+h} - \hat{\mathbf{X}}_T(h) = \sum_{j=0}^{h-1} \mathbf{\Phi}^j \mathbf{a}_{T+h-j}, \tag{3.35}$$

and the mean square error of prediction equals

$$\mathbf{\Sigma}(h) = \text{MSEP}(h) = \sum_{j=0}^{h-1} \mathbf{\Phi}^j \mathbf{\Sigma} (\mathbf{\Phi}^j)'. \tag{3.36}$$

Consider, now, the VAR(p) model with known parameters, an i.i.d. sequence $\mathbf{a}_t$ and $\mathcal{F}_t = \{\mathbf{X}_s : s \leq t\}$. We obtain

$$E(\mathbf{X}_{t+h}|\mathcal{F}_t) = \mathbf{\Phi}_0 + \mathbf{\Phi}_1 E(\mathbf{X}_{t+h-1}|\mathcal{F}_t) + \cdots + \mathbf{\Phi} E(\mathbf{X}_{t+h-p}|\mathcal{F}_t),$$

since $E(\mathbf{a}_{t+h}|\mathcal{F}_t) = 0$, for all $h > 0$.

For $h = 1$,
$$\hat{\mathbf{X}}_t(1) = \mathbf{\Phi}_0 + \mathbf{\Phi}_1 \mathbf{X}_t + \cdots + \mathbf{\Phi}_p \mathbf{X}_{t-p+1},$$

and for $h = 2$

$$\hat{\mathbf{X}}_t(2) = \mathbf{\Phi}_0 + \mathbf{\Phi}_1 \hat{\mathbf{X}}_t(1) + \mathbf{\Phi}_2 \mathbf{X}_t + \cdots + \mathbf{\Phi}_p \mathbf{X}_{t-p+2},$$

therefore, the predictions can be obtained recursively. The prediction error h steps ahead is given by

$$\mathbf{e}_T(h) = \sum_{j=0}^{h-1} \mathbf{\Psi}_j \mathbf{a}_{T+h-j}, \tag{3.37}$$

where the matrices $\mathbf{\Psi}_j$ are obtained recursively by

$$\mathbf{\Psi}_j = \sum_{k=1}^{p-1} \mathbf{\Psi}_{j-k} \mathbf{\Phi}_k, \tag{3.38}$$

with $\mathbf{\Psi}_0 = \mathbf{I}_n$ and $\mathbf{\Phi}_j = 0, j > p$. It follows that the matrix of prediction MSEP(h) becomes

$$\mathbf{\Sigma}(h) = \sum_{j=0}^{h-1} \mathbf{\Psi}_j \mathbf{\Sigma} \mathbf{\Psi}_j'. \tag{3.39}$$

When the parameters of the VAR(p) are estimated, the best predictor of X_{T+h} is now given by

$$\tilde{\mathbf{X}}_T(h) = \hat{\mathbf{\Phi}}_0 + \hat{\mathbf{\Phi}}_1 \tilde{\mathbf{X}}_T(h-1) + \cdots + \hat{\mathbf{\Phi}}_p \tilde{\mathbf{X}}_T(h-p), \quad h > 1. \tag{3.40}$$

In this situation, the MSEP(h) matrix becomes

$$\hat{\boldsymbol{\Sigma}}(h) = \boldsymbol{\Sigma}(h) + \mathrm{MSE}(\mathbf{X}_{T+h} - \tilde{\mathbf{X}}_T(h)). \tag{3.41}$$

In practice, the second term of Equation 3.41 is ignored and $\hat{\boldsymbol{\Sigma}}(h)$ is computed via

$$\hat{\boldsymbol{\Sigma}}(h) = \sum_{j=0}^{h-1} \hat{\boldsymbol{\Psi}}_j \hat{\boldsymbol{\Sigma}} \hat{\boldsymbol{\Psi}}'_j, \tag{3.42}$$

with $\hat{\boldsymbol{\Psi}}_j = \sum_{k=1}^{p-1} \hat{\boldsymbol{\Psi}}_{j-k} \hat{\boldsymbol{\Phi}}_k$. Lütkepohl (1991) gives an approximation for the right-hand side of Equation 3.41.

EXAMPLE 3.6

Let us fit a VAR(p) model to the series $\mathbf{X}_t$ of Example 3.2. Figure 3.3 shows the partial autocorrelation functions for both series, indicating that a model up to order 20 would be feasible. In Table 3.2, we have the values of BIC resulting from the fitting of VAR models up to order 20. According to these values, we select the order $p = 6$. Estimating the model by maximum likelihood, we obtain Table 3.3; hence, the fitted model turns out to be

$$\begin{aligned}
X_{1t} &= 0.204X_{1,t-1} + 0.141X_{2,t-1} + 0.172X_{1,t-2} + 0.116X_{2,t-2} \\
&\quad + 0.113X_{1,t-3} + 0.059X_{1,t-4} + 0.087X_{1,t-5} - 0.076X_{2,t-5} \\
&\quad + 0.158X_{1,t-6} - 0.112X_{2,t-6} + a_{1t}, \tag{3.43} \\
X_{2t} &= 0.124X_{1,t-1} + 0.251X_{2,t-1} + 0.339X_{2,t-2} + 0.193X_{2,t-3} \\
&\quad - 0.101X_{1,t-4} + 0.099X_{2,t-4} + a_{2,t}.
\end{aligned}$$

The constant vector is not significant.

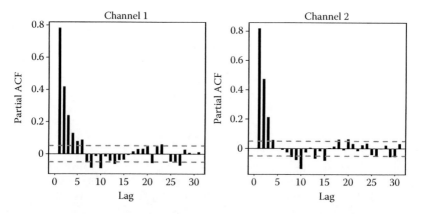

FIGURE 3.3 Sample PACF for the series.

TABLE 3.2
Values of BIC Resulting from Fitting VAR(p) Models,
$p = 1, \ldots, 20$, for Example 3.6

Order	BIC	Order	BIC
1	24,292.46	11	23,710.25
2	23,828.06	12	23,735.18
3	23,737.21	13	23,759.39
4	23,713.21	14	23,786.61
5	23,697.57	15	23,801.17
6	23,669.28*	16	23,820.52
7	23,682.20	17	23,848.44
8	23,704.06	18	23,867.70
9	23,705.64	19	23,895.35
10	23,698.51	20	23,908.29

*Smallest BIC.

TABLE 3.3
Fitting of a VAR(6) Model to X_t of Example 3.8. First Line:
Estimates; Second Line: Standard Deviations

$\hat{\Phi}_1$	$\hat{\Phi}_2$	$\hat{\Phi}_3$
$\begin{bmatrix} 0.204 & 0.141 \\ 0.124 & 0.251 \end{bmatrix}$	$\begin{bmatrix} 0.172 & 0.116 \\ 0.042 & 0.339 \end{bmatrix}$	$\begin{bmatrix} 0.113 & 0.042 \\ 0.055 & 0.193 \end{bmatrix}$
$\begin{bmatrix} 0.028 & 0.028 \\ 0.029 & 0.029 \end{bmatrix}$	$\begin{bmatrix} 0.029 & 0.029 \\ 0.029 & 0.029 \end{bmatrix}$	$\begin{bmatrix} 0.029 & 0.029 \\ 0.030 & 0.030 \end{bmatrix}$

$\hat{\Phi}_4$	$\hat{\Phi}_5$	$\hat{\Phi}_6$
$\begin{bmatrix} 0.059 & -0.101 \\ 0.023 & 0.099 \end{bmatrix}$	$\begin{bmatrix} 0.087 & -0.050 \\ -0.076 & 0.027 \end{bmatrix}$	$\begin{bmatrix} 0.158 & -0.060 \\ -0.112 & 0.020 \end{bmatrix}$
$\begin{bmatrix} 0.029 & 0.029 \\ 0.030 & 0.030 \end{bmatrix}$	$\begin{bmatrix} 0.028 & 0.029 \\ 0.029 & 0.029 \end{bmatrix}$	$\begin{bmatrix} 0.028 & 0.029 \\ 0.028 & 0.029 \end{bmatrix}$

In Table 3.4, we show the simplified representations of the matrix coefficients. We conclude that there are linear relationships between the series: values of X_{1t} are influenced by past values of X_{2t} and vice versa.

In Figure 3.4, we show the sample ACF of residuals and squared residuals. We see that there is the possibility of further improving the model, by introducing moving average terms or considering heteroskedastic models. But this will not be pursued here.

TABLE 3.4

Simplified Representation for the Matrices in Table 3.3

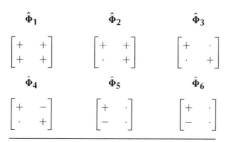

$$\hat{\Phi}_1 \qquad \hat{\Phi}_2 \qquad \hat{\Phi}_3$$

$$\begin{bmatrix} + & + \\ + & + \end{bmatrix} \qquad \begin{bmatrix} + & + \\ \cdot & + \end{bmatrix} \qquad \begin{bmatrix} + & \cdot \\ \cdot & + \end{bmatrix}$$

$$\hat{\Phi}_4 \qquad \hat{\Phi}_5 \qquad \hat{\Phi}_6$$

$$\begin{bmatrix} + & - \\ \cdot & + \end{bmatrix} \qquad \begin{bmatrix} + & \cdot \\ - & \cdot \end{bmatrix} \qquad \begin{bmatrix} + & \cdot \\ - & \cdot \end{bmatrix}$$

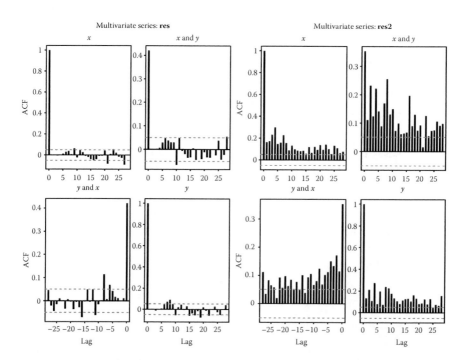

FIGURE 3.4 Sample ACF for the residuals (**res**) and squared residuals (**res2**) of model (3.43).

3.6 GRANGER CAUSALITY

The elucidation of causality relationships between variables is an important topic in empirical research. For temporal systems, Granger (1969) defines causality in terms of predictability: the variable X causes the variable Y, with respect to a given universe of information (which includes X and Y), if the present of Y can be more efficiently predicted using past values of X, all other information being available (including the past of Y). The definition does not require that the system be linear.

Let $\{A_t, t = 0, \pm 1, \pm 2, \ldots\}$ be the relevant set of information up to (and including) t, containing at least X_t, Y_t. Define $\overline{A}_t = \{A_s : s < t\}$, $\overline{\overline{A}}_t = \{A_s : s \leq t\}$, and similar definitions for $\overline{X}_t, \overline{Y}_t$, and so on. Let $P_t(Y|B)$ be the predictor of Y_t with minimum MSE, given the information set B and $\sigma^2(Y|B)$ the corresponding MSE of the predictor.

Definition 3.1   We say that:
   a. $X_t \to Y_t$: X_t *causes* Y_t in the sense of Granger if

$$\sigma^2(Y_t|\overline{A}_t) < \sigma^2(Y_t|\overline{A}_t - \overline{X}_t).$$

This means that Y_t can be better predicted by using all the available information, including the past of Y_t and X_t. We also say that X_t is *exogenous* or *leading* Y_t.
   b. $X_t \Rightarrow Y_t$: X_t *causes instantaneously* Y_t in the sense of Granger if $\sigma^2(Y_t|\overline{A}_t, \overline{\overline{X}}_t) < \sigma^2(Y_t|\overline{A}_t)$, that is, the present value of Y_t is better predicted if the present value of X_t is included.
   c. There is *feedback*, and we write $X_t \leftrightarrow Y_t$, if X_t causes Y_t and Y_t causes X_t.
   d. There is *unidirectional causality* from X_t to Y_t if $X_t \to Y_t$ and *there is no feedback*.

It is easy to see that if $X_t \Rightarrow Y_t$, then $Y_t \Rightarrow X_t$. We simply say that there is instantaneous causality between X_t and Y_t.

The definition extends to random vectors. There are several proposals to make operational the definition of causality. Pierce and Haugh (1977) propose to fit ARIMA models to suitable transformations of both variables and afterward to establish causality patterns between the residuals through cross-correlations. Hsiao (1979) suggests fitting AR models using the AIC criterion. In the case of more than two series, Boudjellaba et al. (1992) suggests fitting VARMA (vector ARMA) models.

In this chapter, we use the VAR representation of the multivariate series $\mathbf{X}_t$, of order $n \times 1$. The MA representation of the process is given by an extension of Equation 3.21, namely,

$$\mathbf{X}_t = \boldsymbol{\mu} + \boldsymbol{\Psi}(B)\mathbf{a}_t, \quad \boldsymbol{\Psi}_0 = \mathbf{I}_n, \tag{3.44}$$

where $\boldsymbol{\Psi}(B) = \mathbf{I}_n + \boldsymbol{\Psi}_1(B) + \boldsymbol{\Psi}_2(B^2) + \cdots$.
   Assume that

$$\mathbf{X}_t = \begin{bmatrix} \mathbf{Y}_t \\ \mathbf{Z}_t \end{bmatrix},$$

where $\mathbf{Y}_t$ is an $r \times 1$ vector and $\mathbf{Z}_t$ is an $s \times 1$ vector, $r + s = n$. Then we can write

$$\mathbf{X}_t = \begin{bmatrix} \mathbf{Y}_t \\ \mathbf{Z}_t \end{bmatrix} = \begin{bmatrix} \boldsymbol{\mu}_1 \\ \boldsymbol{\mu}_2 \end{bmatrix} + \begin{bmatrix} \boldsymbol{\Psi}_{11}(B) & \boldsymbol{\Psi}_{12}(B) \\ \boldsymbol{\Psi}_{21}(B) & \boldsymbol{\Psi}_{22}(B) \end{bmatrix} \begin{bmatrix} \mathbf{a}_{1t} \\ \mathbf{a}_{2t} \end{bmatrix}, \tag{3.45}$$

partitioning $\boldsymbol{\mu}$, $\boldsymbol{\Psi}(B)$, and $\mathbf{a}_t$ according to the partition of $\mathbf{X}_t$. If there is unidirectional causality from $\mathbf{Y}_t$ to $\mathbf{Z}_t$, that is, if $\mathbf{Z}_t$ is better predicted using the past and present of $\mathbf{Y}_t$, but not the opposite, we should have $\boldsymbol{\Psi}_{12}(B) = 0$, and therefore,

$$\mathbf{Y}_t = \boldsymbol{\mu}_1 + \boldsymbol{\Psi}_{11}(B)\mathbf{a}_{1t}, \tag{3.46}$$

$$\mathbf{Z}_t = \boldsymbol{\mu}_2 + \boldsymbol{\Psi}_{21}(B)\mathbf{a}_{1t} + \boldsymbol{\Psi}_{22}(B)\mathbf{a}_{2t}. \tag{3.47}$$

Notice that Equation 3.47 can be written as

$$\mathbf{Z}_t = \boldsymbol{\mu}_2 + V(B)\mathbf{Y}_t + \psi_{22}(B)\mathbf{a}_{2t}, \tag{3.48}$$

which is a dynamical regression model. The conditions $\boldsymbol{\Psi}_{12}(B) = 0$ and $V(B) = 0$ imply that $\mathbf{Y}_t$ does not cause $\mathbf{Z}_t$ and vice versa. In this situation, both series are not coupled or are related only instantaneously. As a matter of fact, it is possible to prove the following result, which is a characterization of Granger noncausality. Note that

$$\boldsymbol{\Psi}_i = \begin{bmatrix} \boldsymbol{\Psi}_{11,i} & \boldsymbol{\Psi}_{12,i} \\ \boldsymbol{\Psi}_{21,i} & \boldsymbol{\Psi}_{22,1} \end{bmatrix}, \quad i = 1, 2, \dots.$$

Proposition 3.4   The optimal predictor of $\mathbf{Y}_t$ based on $\overline{\overline{\mathbf{Z}}}_t$ is the same as the optimal predictor of $\mathbf{Y}_t$ based on $\overline{\overline{\mathbf{Y}}}_t$ if and only if $\boldsymbol{\Psi}_{12,i} = \mathbf{0}$, $i = 1, 2, \dots$.

In other words, $\mathbf{Z}_t$ does not cause $\mathbf{Y}_t$ if and only if $\boldsymbol{\Psi}_{12,i}(B) = \mathbf{0}$, for all $i \geq 1$. For proof, see Lütkepohl (1991). This proposition applies not only to VAR models, but also to any process that can be written in the form of an infinite MA, and gives a way to verify the nonexistence of Granger causality. Under a practical point of view, it is convenient to consider a finite order VAR model, namely

$$\mathbf{X}_t = \begin{bmatrix} \mathbf{Y}_t \\ \mathbf{Z}_t \end{bmatrix} = \begin{bmatrix} \boldsymbol{\mu}_1 \\ \boldsymbol{\mu}_2 \end{bmatrix} + \begin{bmatrix} \boldsymbol{\Phi}_{11,1} & \boldsymbol{\Phi}_{12,1} \\ \boldsymbol{\Phi}_{21,1} & \boldsymbol{\Phi}_{22,1} \end{bmatrix} \begin{bmatrix} \mathbf{Y}_{t-1} \\ \mathbf{Z}_{t-1} \end{bmatrix} + \cdots$$

$$+ \begin{bmatrix} \boldsymbol{\Phi}_{11,p} & \boldsymbol{\Phi}_{12,p} \\ \boldsymbol{\Phi}_{21,p} & \boldsymbol{\Phi}_{22,p} \end{bmatrix} \begin{bmatrix} \mathbf{Y}_{t-p} \\ \mathbf{Z}_{t-p} \end{bmatrix} + \begin{bmatrix} \mathbf{a}_{1t} \\ \mathbf{a}_{2t} \end{bmatrix}, \tag{3.49}$$

and the condition of Proposition 3.4 is satisfied if and only if $\boldsymbol{\Phi}_{12,i} = \mathbf{0}$, $i = 1, 2, \dots, p$. Or if $\mathbf{X}_t$ follows a VAR(p) model, with nonsingular covariance matrix, then $\mathbf{Z}_t$ does not cause $\mathbf{Y}_t$ if and only if $\boldsymbol{\Phi}_{12,i} = \mathbf{0}$, $i = 1, 2, \dots, p$.

A characterization of nonexistence of instantaneous causality is given by the following result (Lütkepohl, 1991).

Proposition 3.5   If $\mathbf{X}_t$ is as in Equation 3.49, with a nonsingular covariance matrix, then there is no instantaneous Granger causality between $\mathbf{Y}_t$ and $\mathbf{Z}_t$ if and only if $E(\mathbf{a}_{1t}\mathbf{a}_{2t}') = 0$.

The proof of the proposition is based on the fact that the covariance matrix $\boldsymbol{\Sigma}$ can be written as $\boldsymbol{\Sigma} = \mathbf{T}\mathbf{T}'$, where $\mathbf{T}$ is a lower triangular matrix, with positive elements in the main diagonal. Then, the MA representation of $\mathbf{X}_t$ can be written as

$$\mathbf{X}_t = \boldsymbol{\mu} + \sum_{j=0}^{\infty} \boldsymbol{\Psi}_j \mathbf{T} \mathbf{T}^{-1} \mathbf{a}_{t-j}$$

$$= \boldsymbol{\mu} + \sum_{j=0}^{\infty} \boldsymbol{\Xi}_j \mathbf{b}_{t-j},$$

with $\boldsymbol{\Xi}_j = \boldsymbol{\Psi}_j \mathbf{T}$, $\mathbf{b}_t = \mathbf{T}^{-1} \mathbf{a}_t \sim \mathrm{WN}(\mathbf{0}, \boldsymbol{\Sigma}_b)$, and $\boldsymbol{\Sigma}_b = \mathbf{T}^{-1} \boldsymbol{\Sigma} (\mathbf{T}^{-1})' = \mathbf{I}_n$.
Equation 3.49 can be written as

$$\mathbf{Y}_t = \boldsymbol{\mu}_1 + \sum_{i=1}^{p} \boldsymbol{\Phi}_{11,i} \mathbf{Y}_{t-i} + \sum_{i=1}^{p} \boldsymbol{\Phi}_{12,i} \mathbf{Z}_{t-i} + \mathbf{a}_{1t}, \tag{3.50}$$

$$\mathbf{Z}_t = \boldsymbol{\mu}_2 + \sum_{i=1}^{p} \boldsymbol{\Phi}_{21,i} \mathbf{Y}_{t-i} + \sum_{i=i}^{p} \boldsymbol{\Phi}_{22,i} \mathbf{Z}_{t-i} + \mathbf{a}_{2t}. \tag{3.51}$$

Suppose, also, that $\boldsymbol{\Sigma}$ is partitioned as

$$\boldsymbol{\Sigma} = \begin{bmatrix} \boldsymbol{\Sigma}_{11} & \boldsymbol{\Sigma}_{12} \\ \boldsymbol{\Sigma}_{21} & \boldsymbol{\Sigma}_{22} \end{bmatrix},$$

with $\boldsymbol{\Sigma}_{ij} = E(\mathbf{a}_{it} \mathbf{a}_{jt}')$, $i, j = 1, 2$.
Then, as we saw above,

   i. $\mathbf{Z}_t$ does not cause $\mathbf{Y}_t \leftrightarrow \boldsymbol{\Phi}_{12,i} = \mathbf{0}$, for all i;
  ii. $\mathbf{Y}_t$ does not cause $\mathbf{Z}_t \leftrightarrow \boldsymbol{\Phi}_{21,i} = \mathbf{0}$, for all i.

Equivalent results to (i) and (ii) are given in the following proposition.

Proposition 3.6

   i. $\mathbf{Z}_t$ does not cause $\mathbf{Y}_t \leftrightarrow |\boldsymbol{\Sigma}_{11}| = |\boldsymbol{\Sigma}_1|$, where $\boldsymbol{\Sigma}_1 = E(\mathbf{c}_{1t} \mathbf{c}_{1t}')$ is obtained from the restricted regression

$$\mathbf{Y}_t = \boldsymbol{v}_1 + \sum_{i=1}^{p} \mathbf{A}_i \mathbf{Y}_{t-i} + \mathbf{c}_{1t}. \tag{3.52}$$

  ii. $\mathbf{Y}_t$ does not cause $\mathbf{Z}_t \leftrightarrow |\boldsymbol{\Sigma}_{22}| = |\boldsymbol{\Sigma}_2|$, where $\boldsymbol{\Sigma}_2 = E(\mathbf{c}_{2t} \mathbf{c}_{2t}')$ is obtained from the restricted regression

$$\mathbf{Z}_t = \boldsymbol{v}_2 + \sum_{i=1}^{p} \mathbf{C}_i \mathbf{Z}_{t-i} + \mathbf{c}_{2t}. \tag{3.53}$$

The regressions (3.50) through (3.53) can be estimated via OLS and from the residuals of these regressions, we get the estimators

$$\hat{\boldsymbol{\Sigma}}_i = (T-p)^{-1} \sum_{t=p+1}^{T} \hat{c}_{it}\hat{c}'_{it},$$

$$\hat{\boldsymbol{\Sigma}}_{ii} = (T-p)^{-1} \sum_{t=p+1}^{T} \hat{a}_{it}\hat{a}'_{it}, \quad i = 1, 2.$$

The tests and corresponding likelihood ratio statistics are given by

i. H_{01} : $\boldsymbol{\Phi}_{12,i} = \mathbf{0}$, for all i ($\mathbf{Z}_t$ does not cause $\mathbf{Y}_t$),

$$RV_1 = (T-p)[\log|\hat{\boldsymbol{\Sigma}}_1| - \log|\hat{\boldsymbol{\Sigma}}_{11}|] \sim \chi^2(prs).$$

ii. H_{02} : $\boldsymbol{\Phi}_{21,i} = \mathbf{0}$, for all i ($\mathbf{Y}_t$ does not cause $\mathbf{Z}_t$),

$$RV_2 = (T-p)[\log|\hat{\boldsymbol{\Sigma}}_2| - \log|\hat{\boldsymbol{\Sigma}}_{22}|] \sim \chi^2(prs).$$

EXAMPLE 3.7

For the Example 3.6, we see that $X_{1t} \to X_{2t}$, and $X_{2t} \to X_{1t}$, that is, there is a feedback relationship between the series.

3.7 FURTHER REMARKS

VAR models have been used often in neuroscience, for example, in problems related to connectivity in fMRI. See Goebel et al. (2003) for details and references. Also, often series appearing in fMRI are not stationary and some form of time-varying coefficient models are needed. See Sato et al. (2006a,b) for dynamic VAR models using wavelets. There is a connection, between VAR models, Granger causality and the concept of partial directed coherence. See Baccalá and Sameshima (2001) and Takahashi et al. (2007) for further details (and also Chapters 4 and 7).

REFERENCES

Baccalá, L. A. and K. Sameshima. 2001. Partial directed coherence: A new concept in neural structure determination. *Biol. Cybern. 84*: 463–474.

Boudjellaba, H., J.-M. Dufour, and R. Roy. 1992. Testing causality between two vectors in multivariate autoregressive moving average models. *J. Am. Stat. Assoc. 87*: 1082–1090.

Brockwell, P. J. and R. A. Davis. 1991. *Time Series: Theory and Methods*. 2nd Ed. New York: Springer.

Fuller, W. A. 1996. *Introduction to Statistical Time Series*. 2nd Ed. New York: Wiley.

Goebel, R., A. Roebroeck, D. S. Kim et al. 2003. Investigating directed cortical interactions in time-resolved fMRI data using vector autoregressive modelling and Granger causality mapping. *Magn. Reson. Imag. 21*: 1251–1261.

Granger, C. W. J. 1969. Investigating causal relationships by econometric models and cross-spectral methods. *Econometrica 37*: 424–438.

Hsiao, C. 1979. Autoregressive modelling of Canadian money and income data. *J. Am. Stat. Assoc. 74*: 553–560.

Lütkepohl, H. 1991. *Introduction to Multiple Time Series Analysis*. Heidelberg: Springer-Verlag.

Pierce, D. A. and L. D. Haugh. 1977. Causality in temporal systems: Characterizations and a survey. *J. Econom. 5*: 265–293.

Sato, J. R., P. A. Morettin, P. R. Arantes et al. 2006a. Wavelet based time-varying vector autoregressive modelling. *Comput. Stat. Data An. 51*: 5847–5866.

Sato, J. R., M. M. Felix, E. Amaro Jr et al. 2006b. A method to produce evolving functional connectivity maps during the course of an fMRI experiment using wavelet-based time-varying Granger causality. *NeuroImage 31*: 187–196.

Takahashi, D. Y., L. A. Baccalá, and K. Sameshima. 2007. Connectivity inference via partial directed coherence: Asymptotic results. *J. Appl. Stat. 34*: 1259–1273.

Tsay, R. S. 2005. *Analysis of Financial Time Series*. 2nd Ed. New York: Wiley.

4 Partial Directed Coherence

Luiz A. Baccalá and Koichi Sameshima

CONTENTS

4.1 INTRODUCTION

The search for brain structures that might be responsible for driving typical EEG rhythms has been recurrent in neuroscience. For many decades, the use of pairwise signal correlation has been this quest's tool of choice, possibly because it is easy to understand and compute, especially under the prevailing technological limitations before the 1990s personal computer's climb to ubiquity.

Under such a rapidly evolving scenario of ever-increasing computational power and the feasibility of affordable multichannel neurophysiologic recording (Eichenbaum and Davis, 1998; Nicolelis, 1998), it was soon becoming apparent that correlation-based analyses were fraught with limitations that were not easy to overcome and that often led to contradictory interpretations (Baccalá and Sameshima, 2001c).

An interesting option seemed to lie in the systematic adoption of time-series analysis methods, of which Gersch's proposal is an early example (Gersch, 1970a,b), but whose popularity was surely limited by its early 1970s appearance long before the personal computer revolution.

The first paper to attract our attention in this regard was Schnider et al. (1989), who looked for plausible mechanisms behind the tremors in Parkinson's disease in the form of absent or defective signal feedback in the brain/muscle signal pathway. The paper was interesting on several counts: (a) it made use of system identification considerations that allowed the problem to be stated in the frequency domain, which is useful for characterizing oscillations, and (b) it made fundamental reference to an idea in econometrics associated with Clive Granger, which came to be known as *Granger causality* (or G-causality).

Granger causality (Granger, 1969) has been the subject of many investigations, especially in the time domain, given its original aim of characterizing issues such as pointing out the possible "culprits" behind the empirical behavior of economic variables such as interest rates, unemployment, and GDP (gross domestic product). Its core property is that it allows time precedence to be exposed between dynamically evolving quantities, which can be interpreted as an indication of how information flows from one observed variable to another, thus leading to what came to be known as *directed* interaction.

While Gersch's contribution was centered on the idea of *partial coherence* and applied to sets of three simultaneously acquired time series (Gersch, 1972), some other important early contributions were more specific at attempting frequency domain characterizations of Granger causality (Geweke et al., 1983; Geweke, 1984; and notably Saito and Harashima, 1981) but at the expense of examining just pairs of dynamical variables at a time, thus suffering from the same interpretation problems that limit pairwise time-series correlation analysis (Baccalá and Sameshima, 2001c).

In 1991, Kamiński and Blinowska introduced directed transfer function (DTF) (Chapter 2) (Kamiński and Blinowska, 1991).

Our investigation began as an attempt to appraise the relationship between (Kamiński and Blinowska, 1991) with what was becoming standard in testing Granger causality in a multivariate time-series context—Wald-type model likelihood tests (Lütkepohl, 1993). The results were reported in Baccalá et al. (1998), and exposed contradictory conclusions regarding connectivity in some cases. Further discussion of the reason is best left to Section 4.2. For now, it is sufficient to say that the breakthrough came when we realized that the *partial coherence* between time series could be written in a simple way using multivariate autoregressive models (Baccalá, 2001) and that its factorization in a form analogous to what Kamiński and Blinowska (1991) had done with *coherence* produced a new connectivity estimator that we named *partial directed coherence* (PDC), which now coincided with the time-domain G-causality tests as described in Lütkepohl (1993).

4.2 MATHEMATICAL FORMULATION

The first step toward understanding PDC is to realize that a simple way to describe the joint dynamic behavior of multivariate time series is through *vector autoregressive* models:

$$
\begin{bmatrix} x_1(n) \\ \vdots \\ x_N(n) \end{bmatrix} = \sum_r \mathbf{A}_r \begin{bmatrix} x_1(n-r) \\ \vdots \\ x_N(n-r) \end{bmatrix} + \begin{bmatrix} w_1(n) \\ \vdots \\ w_N(n) \end{bmatrix} \tag{4.1}
$$

whose cofficient matrix for the r-th past lag

$$
\mathbf{A}_r = \begin{bmatrix} a_{11}(r) & a_{12}(r) & \cdots & \cdots & a_{1N}(r) \\ \vdots & \vdots & \vdots & \vdots & \vdots \\ \vdots & \vdots & \vdots & a_{ij}(r) & \vdots \\ \vdots & \vdots & \vdots & \vdots & \vdots \\ a_{N1}(r) & \cdots & \cdots & \cdots & a_{NN}(r) \end{bmatrix}
$$

is formed by the $a_{ij}(r)$ coefficients which represent the linear interaction effect of $x_j(n-r)$ onto $x_i(n)$ and where $w_i(n)$ represent *innovation* stochastic processes, that is, that part of the dynamical behavior that cannot be predicted from past observations of the processes. The $w_i(n)$ are thus uncorrelated over time but may exhibit instantaneous correlations among themselves which are described by its covariance matrix

$$
\mathbf{\Sigma_w} = \begin{bmatrix} \sigma_{11}^2 & \sigma_{12} & \cdots & \cdots & \sigma_{1N} \\ \vdots & \vdots & \vdots & \vdots & \vdots \\ \vdots & \vdots & \vdots & \sigma_{ij} & \vdots \\ \vdots & \vdots & \vdots & \vdots & \vdots \\ \sigma_{N1} & \cdots & \cdots & \cdots & \sigma_{NN}^2 \end{bmatrix}, \tag{4.2}
$$

where $\sigma_{ij} = cov(w_i(n), w_j(n))$.

Equation 4.1 summarizes a first linear approximation to describe the interactions between the $x_i(n)$ time series. The $a_{ij}(r)$ coefficients may be obtained by adequate minimization of the mean-squared prediction error:

$$
E\left[|\mathbf{x}(n) - \hat{\mathbf{x}}(n)|^2\right], \tag{4.3}
$$

where $\mathbf{x}(n) = [x_1(n), \ldots, x_N(n)]^T$ are the observations and

$$
\hat{\mathbf{x}}(n) = \sum_{r=1}^{p} \mathbf{A}_r \mathbf{x}(n-r), \tag{4.4}
$$

the linear predictor of $\mathbf{x}(n)$, given its past (in fact, up to p lags into the past where p is termed the *model order*).

There are a variety of ways for minimizing Equation 4.3 leading to a *unique* estimate for $a_{ij}(r)$.

By far, the most popular use for adjusting Equation 4.1 is time-series prediction, but it can also be used for multivariate parametric spectral estimation (Chapter 2) with the help of the model's frequency domain representation:

$$\mathbf{A}(\lambda) = \left. \sum_{r=1}^{p} \mathbf{A}_r z^{-r} \right|_{z=e^{-j2\pi\lambda}}, \tag{4.5}$$

where $\mathbf{j} = \sqrt{-1}$ and $|\lambda| \leq 0.5$ is the normalized frequency so that $\lambda = 0.5$ represents one half the time series sampling rate f_s.*

The breakthrough idea came from realizing that Equation 4.5 could be used to write *partial coherence* in a simple way (Baccalá, 2001):

$$\kappa_{ij}^2(\lambda) = \frac{\bar{a}_i^H(\lambda)\boldsymbol{\Sigma}_{\mathbf{w}}^{-1}\bar{a}_j(\lambda)}{\sqrt{\left(\bar{a}_i^H(\lambda)\boldsymbol{\Sigma}_{\mathbf{w}}^{-1}\bar{a}_i(\lambda)\right)\left(\bar{a}_j^H(\lambda)\boldsymbol{\Sigma}_{\mathbf{w}}^{-1}\bar{a}_j(\lambda)\right)}}, \tag{4.6}$$

where $\bar{a}_i(\lambda)$ are the columns of the $\bar{\mathbf{A}}(\lambda) = \mathbf{I} - \mathbf{A}(\lambda)$ matrix.

Abstracting the use of $\boldsymbol{\Sigma}_{\mathbf{w}}$ and factorizing Equation 4.6 in a way similar to that done for DTF (Chapter 2) led to the original definition of PDC from $x_j(n)$ to $x_i(n)$:

$$\pi_{ij}(\lambda) = \frac{\bar{A}_{ij}(\lambda)}{\sqrt{\sum_{k=1}^{N}|\bar{A}_{kj}(\lambda)|^2}} = \frac{\bar{A}_{ij}(\lambda)}{\sqrt{\bar{a}_j^H(\lambda)\bar{a}_j(\lambda)}}, \tag{4.7}$$

where $\bar{A}_{ij}(\lambda)$ is the i,j-th element of $\bar{\mathbf{A}}(\lambda)$.

It is important to note that

$$\bar{A}_{ij}(\lambda) = \begin{cases} 1 - \displaystyle\sum_{r=1}^{p} a_{ij}(r)e^{-j2\pi\lambda r}, & \text{if } i = j \\[3mm] -\displaystyle\sum_{r=1}^{p} a_{ij}(r)e^{-j2\pi\lambda r}, & \text{otherwise} \end{cases} \tag{4.8}$$

so that $\pi_{ij}(\lambda)$'s nullity is governed by the nullity of $a_{ij}(r)$ coefficients, which summarize the influence of $x_j(n-r)$ past observations ($r = 1,\ldots,p$) on the observed variable, $x_i(n)$.

---

* The normalized scale needs to be multiplied by f_s to obtain the actual physical frequency of interest. Let $T_s = 1/f_s$ represent the sampling interval.

Because of its definition as a ratio in Equation 4.7, the following normalization properties hold:

$$0 \leq |\pi_{ij}(\lambda)|^2 \leq 1 \qquad (4.9)$$

$$\sum_{i=1}^{N} |\pi_{ij}(\lambda)|^2 = 1. \qquad (4.10)$$

$\boldsymbol{\Sigma_w}$ is absent from Equation 4.7 even though Equation (15) in Baccalá and Sameshima (2001a) utilizes it as a PDCF—*partial directed coherence factor* and then simplifies it to the form in Equation 4.7. It may be interesting to note that this legacy form is due to an early submission of Baccalá and Sameshima (2001a), which was rejected without comments by another journal where we tried to formulate PDC within a more general framework of "metrics" involving columns of $\bar{\boldsymbol{A}}(\lambda)$. The "metric" nomenclature still survives in the software package that accompanies this book and proved important in the development of Section 4.2.1 and in connection to information PDC (*ιPDC*) as discussed in Chapter 5. To simplify matters, we included $\boldsymbol{\Sigma_w}$ only marginally in Baccalá and Sameshima (2001a), a lucky decision, since its full significance was only to became clear in Takahashi et al. (2010) (see also Chapter 5).

In the meantime, while applying PDC to various scenarios, we came across the Melanoma/Sunspot bivariate time series examined by Andrews and Herzberg (1985) who provided yearly data on the incidence of melanomas in Connecticut and related it to solar activity as measured by the number of observed Sunspots over the same period of 37 years. Physically, only the sunspot to melanoma directed causal relation is possible. Yet computing Equation 4.7, as originally defined, apparently led to the opposite absurd conclusion (see Figure 4.1a) when indeed G-causality tests as described in Chapter 3 were in accord with the known physical mechanisms. Application of DTF as defined in Kamiński and Blinowska (1991) produced the same counterintuitive results.*

Though we have never seen this suggestion in print, an *ad hoc* bypass to this problem came from standardizing the time series prior to modeling, that is the series have their means subtracted and are divided by their standard deviations. This produces dimensionless time series, which are then subject to modeling and further PDC computation. Doing this does indeed apparently correct the problem (Figure 4.1b).

We wondered whether a more general and elegant solution existed. Drawing from the dimensionality clue, we realized that the originally defined PDC is not scale invariant, that is, arbitrary changes in the amplitudes of one time series can lead to substantial changes in PDC values. This deficiency was addressed by introducing the notion of *generalized* PDC (*g*PDC) discussed next.

---

* For bivariate time series, DTF and PDC are indistinguishable except for numerical accuracy issues.

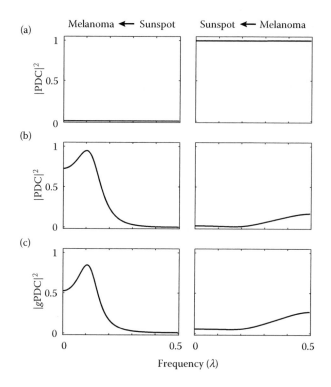

FIGURE 4.1 *Sunspot* → *Melanoma* (left panels) versus *Melanoma* → *Sunspot* (right panels) squared magnitude PDC (a) without and (b) with data standardization complemented by (c) gPDC, showing that intuitive causal results are obtained by gPDC without the need to resort to data standardization. Refer to Example 7.5 to realize that correct connectivity detection inference is possible even in case (a), if correct threshold levels are employed. These graphs are edited versions from **sunspot_melanoma_pdc_gdc.m**.

4.2.1 Generalized PDC

To make PDC scale invariant, we redefined it (Baccalá et al., 2007) as

$$
{}_g\pi_{ij}(\lambda) = \frac{\dfrac{1}{\sigma_{ii}}\bar{A}_{ij}(\lambda)}{\sqrt{\displaystyle\sum_{k=1}^{N}\frac{1}{\sigma_{kk}^2}|\bar{A}_{kj}(\lambda)|^2}}. \tag{4.11}
$$

This not only solved the Melanoma/Sunspot paradox (Figure 4.1c) but also proved statistically superior to signal standardization practices (Baccalá et al., 2007). The type of normalization properties that hold for PDC holds for gPDC as well, that is, $|{}_g\pi_{ij}(\lambda)|^2$ is from 0 to 1 and summing it over the i (signal receiving target) adds to 1.

Finally, gPDC reduces to PDC whenever the innovation variances σ_{ii}^2 are all identical.

It is perhaps worth remarking at this point that when gPDC was introduced, very little was known about the statistical behavior of PDC, except that it was frequency dependent (Baccalá and Sameshima, 2001b). With the development of the appropriate statistical procedures, it is now clear that PDC and gPDC are inferentially identical, provided that the appropriate decision threshold to each case is employed (see Chapter 7).

4.3 NUMERICAL ILLUSTRATIONS

Regardless of the approach, a critical step in connectivity computations is to produce an adequate time-series model. This is done via well-defined procedures that comprise model choice, which in our case reduces to choosing the order p of the vector autoregressive (VAR) model to fit. Lastly, before computing PDC, it is important to assert model quality. This involves checking the model prediction residuals for whiteness (Chapter 3). Despite its crucial nature when dealing with real data, as illustrated in some detail for the Sunspot/Melanoma data fitted in Section 7.3.2, these steps are often left implicit in the technical literature and represent a major stumbling block in practical inference.

4.3.1 SYNTHETIC MODELS

Synthetic data are used in the following two examples: (a) a simple bivariate problem (Example 4.1) is used to explore the effect of distinct channel innovation covariances, and (b) the canonical example involving $N = 5$ time series in Baccalá and Sameshima (2001a) (Example 4.3).

The examples are illustrated in what we called the *standard form*: the results are displayed in a panel containing plots organized according to a matrix-like display where rows represent the time series being influenced and begin from $i = 1$ at the top, while columns refer to sources that begin at $j = 1$ from the left. Because of the normalization Property 4.2 (4.10), it has become customary to display the time series ($i = j$) power spectra along the matrix panel diagonal (shaded gray); this is important and serves as a further means of checking the joint model quality, since the spectra of each time series may be estimated independently.

EXAMPLE 4.1

Consider the simplest possible case of directed connection as represented by the model:

$$\begin{cases} x_1(n) = 0.95\sqrt{2}x_1(n-1) - 0.9025x_1(n-2) + w_1(n) \\ x_2(n) = -0.5x_2(n-1) + 0.5x_1(n-1) + w_2(n) \end{cases} \quad (4.12)$$

where $w_i(k)$ are i.i.d. mutually independent random zero mean unit variance Gaussian sequences. The imposed connectivity between $x_1(k)$ and $x_2(k)$ is summarized in Figure 4.2.

FIGURE 4.2 The connectivity pattern for Example 4.1 is the simplest possible graphical summary of the relationship between the time series.

One thousand data points were generated, and only the last 200 points were used in fitting the model via least squares leading to the PDC results plotted in Figure 4.3.   □

EXAMPLE 4.2

Again, using the same model (4.12) (and the same data), as before, but with $w_i(n)$ variance imbalances $\sigma_{11}^2/\sigma_{22}^2 = 10,000$ reproduces the problem reported for the Melanoma-Sunspot scenario of Section 4.2.1. Compare Figures 4.4 and 4.5, which respectively portray PDC versus gPDC, vindicating the correctness of the latter.   □

The next example, used by many, often without reference, was introduced in Baccalá and Sameshima (2001a) to illustrate the differences between PDC and DTF; next is a slightly modified form: the feedback adopted here is from $x_5(n)$ to $x_2(n)$ instead of to $x_1(n)$ as in Baccalá and Sameshima (2001a).

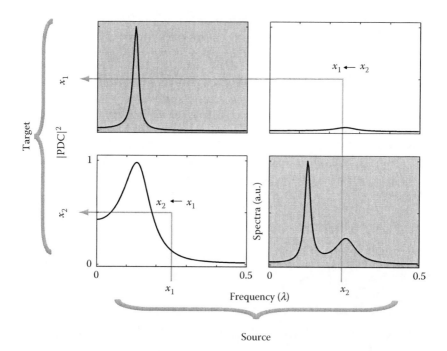

FIGURE 4.3 PDC plot for Example 4.1 presented in *standard form* display, created with **trivial_pdc.m**, showing a high $|PDC|^2$ for $x_1 \rightarrow x_2$ peak around $\lambda = 0.15$ yet being low in the opposite direction.

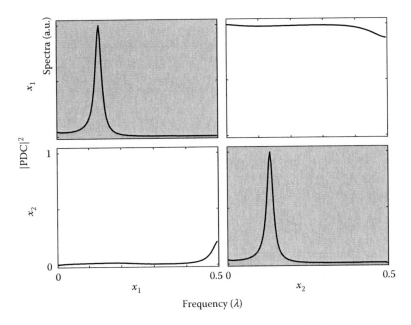

FIGURE 4.4 Standard form PDC display for the unbalanced innovations in Example 4.2 (see **trivial_pdc2.m** routine). Compare it with the *g*PDC results in Figure 4.5.

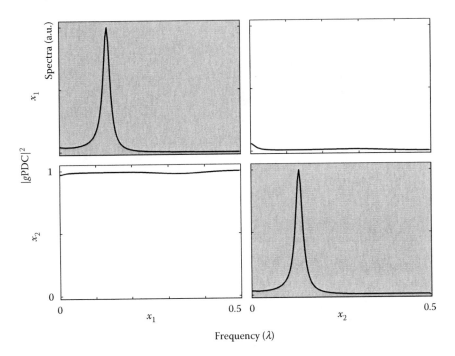

FIGURE 4.5 *g*PDC results on unbalanced innovation channels from Example 4.2 (see **trivial_gpdc.m**). Note that *g*PDC from $x_1(n) \rightarrow x_2(n)$ is flat and consistent with the fact that $x_1(n)$ appears just as a delayed attenuated factor in Equation 4.12 influencing $x_2(n)$.

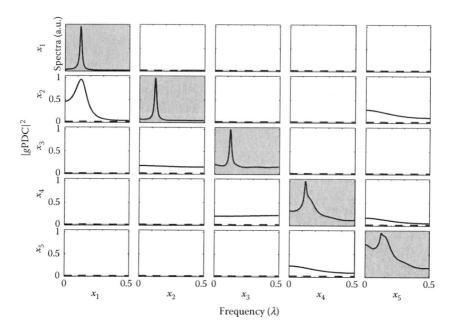

FIGURE 4.6 Since the simulation of Equation 4.13 was performed for identical unit innovations variance (as can be seen running **baccala2001a_ex5m.m** routine), gPDC produces the same results of PDC (cf. Example 5 of Baccalá and Sameshima 2001a). The present connectivity structure is portrayed in the graph of Figure 4.10a.

EXAMPLE 4.3

The dynamics of $N = 5$ time series was designed to consist of a pair of linear oscillators: the first one is represented by $x_1(n)$ while the second is explicitly constructed by the mutual feedback between $x_4(k)$ and $x_5(n)$. The proper frequencies of both oscillators were tuned to be identical in the absence of connections between them:

$$\begin{cases} x_1(n) = 0.95\sqrt{2}x_1(n-1) - 0.9025x_1(n-2) + w_1(n) \\ x_2(n) = -0.5x_1(n-1) + 0.5x_5(n-2) + w_2(n) \\ x_3(n) = 0.4x_2(n-2) + w_3(n) \\ x_4(n) = -0.5x_3(n-1) + 0.25\sqrt{2}x_4(n-1) + 0.25\sqrt{2}x_5(n-1) + w_4(n) \\ x_5(n) = -0.25\sqrt{2}x_4(n-1) + 0.25\sqrt{2}x_5(n-1) + w_5(n) \end{cases}$$

The routine implementing the example is in the AsympPDC package provided in this book: **baccala2001a_ex5m.m**. The standard display form connectivity results are confirmed in Figure 4.6. □

4.3.2   REAL DATA

Next, we consider an example of the procedure for dealing with real data.

Consider the Sunspot/Melanoma data previewed in Section 4.2. Its time evolution is shown in Figures 4.7a and 4.7b. Note how the Melanoma incidence has been

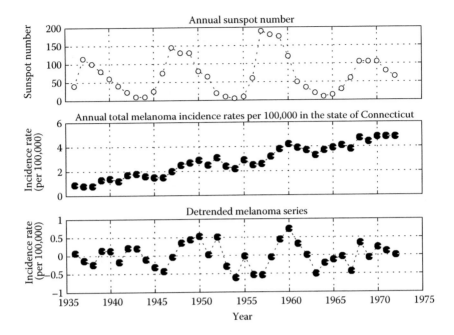

FIGURE 4.7 Comparison plots for the Sunspot-Melanoma data, from Andrews and Herzberg (1985). Both this figure and Figure 4.8 can be obtained from **sunspot_melanoma_series.m**.

increasing steadily and showing a trend. The first step for modeling is to remove that trend. This is done by subtracting a least-squares fitted line to the data resulting in Figure 4.7c, whose residuals are then subject to analysis.

Even if predictive time-series modeling is the ultimate goal, the computation of pairwise cross-correlation functions is always advisable; its results are shown in Figure 4.8 for the case revealing a peak around a lag, which signals a delay of 2 years in their mutual relationship, a fact that already indicates the underlying directed relationship.

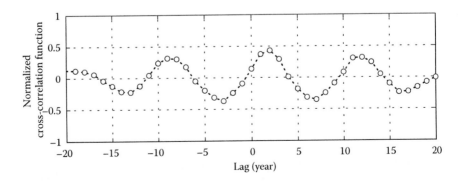

FIGURE 4.8 Sunspot-Melanoma cross-correlation function having a maxium at a 2-year lag pointing to the actual causality from solar activity to melanoma incidence.

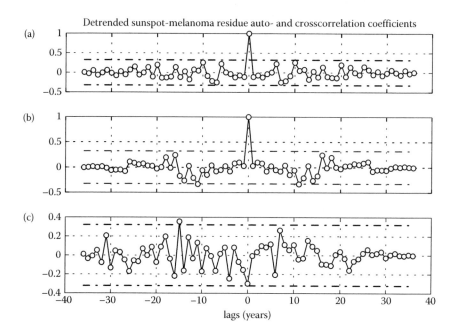

FIGURE 4.9 Model residue autocorrelation Sunspot (a) and Melanoma (b), and (c) cross-correlation function results showing model consistency since residue whiteness cannot be rejected as the nonlag zero coefficients lie within the $\pm 1.96/\sqrt{37}$ limits represented by the dashed lines. See routine **sunspot_melanoma_residue.m** for residue analysis.

Model diagnostics in Figure 4.9 are in favor of model acceptance since residue whiteness cannot be rejected as auto/cross-correlation coefficients among lagged model residues are within the 95% confidence interval for the number of observed data points (see also Chapter 3).

PDC/gPDC and PDC on pre-standardized data are shown in Figure 4.1. See Chapter 7 for results that include statistical tests on this same data set.

4.4 PDC AND OTHER CONNECTIVITY MEASURES

Partial directed coherence is not the only recently directed connectivity measure to have attracted attention. Its relationship to some alternative proposals is briefly outlined next.

4.4.1 PDC AND GRANGER CAUSALITY

Granger causality is gauged by comparing the impact that explicitly taking into consideration the past of one time series has on the predictability of another time series (see Chapter 3). Hence, inferring G-causality absence is based on comparing the model likelihoods between the full free parameter prediction model (4.3) with that of a model fitted under the restriction of nullity for the parameters of interest.

Thus, to test lack of causality from j to i, one must impose the condition:

$$a_{ij}(r) = 0, \quad \forall r = 1, \ldots, p \tag{4.13}$$

in the model fitting procedure and verify its effect comparative to the model likelihood ratio (Lütkepohl, 2005) whose parameters are free. These are known as Wald likelihood ratio tests (see also Baccalá et al. 1998 for other details and the routine **gct_alg.m** that implements the test and is provided with the electronic supplement to this volume).

Therefore, since PDC/gPDC depend directly on Equation 4.13, look at Equations 4.7 and 4.11. This type of test furnishes results of the same kind as PDC and this qualifies PDC/gPDC as a frequency domain portrait of G-causality.

4.4.2 PDC AND DTF

The directed transfer function was discussed at length in Chapter 2. It is built using the rows of the inverse of $\bar{\mathbf{A}}(\lambda)$, that is,

$$\mathbf{H}(\lambda) = \bar{\mathbf{A}}^{-1}(\lambda) \tag{4.14}$$

so that

$$DTF_{ij}(\lambda) = \frac{H_{ij}(\lambda)}{\sum_{k=1}^{N} |H_{ik}(\lambda)|^2}. \tag{4.15}$$

While PDC captures the direct influence of $x_j(n)$ onto $x_i(n)$ discounting the influence of all other intervening time series by factoring them out, DTF measures the net influence of $x_j(n)$ onto $x_i(n)$, taking into account all signal transmission pathways from $x_j(n)$ onto $x_i(n)$.

In Baccalá et al. (1998), we examined the role of *directed coherence* (DC),

$$\gamma_{ij}(\lambda) = \frac{\sigma_{jj} H_{ij}(\lambda)}{\sum_{k=1}^{N} \sigma_{kk}^2 |H_{ik}(\lambda)|^2}, \tag{4.16}$$

which plays the same factorization role with respect to DTF that gPDC plays with regard to PDC (Baccalá et al., 2007).

Similar to PDC, which can be interpreted as a factorization of partial coherence (4.6), DTF can be shown (Baccalá and Sameshima, 2001a) to play the same role for ordinary coherence:

$$|C_{ij}(\lambda)|^2 = \frac{|S_{ij}(\lambda)|^2}{S_{ii}(\lambda) S_{jj}(\lambda)}, \tag{4.17}$$

where $S_{ij}(\lambda)$ and $S_{ii}(\lambda)$ stand for the cross and auto spectra of the time series, respectively.

Both PDC and DTF, in all guises, are complementary measures and answer different aspects of the connectivity problem. Whereas PDC addresses the immediate direct

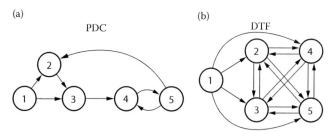

FIGURE 4.10 Comparison between PDC's connectivity disclosure (a) versus DTF's reachability representation (b) for Example 4.3.

dynamic frequency domain link between the various time series (see Figure 4.10a for Example 4.3), DTF exposes the links from one time series to another regardless of the influence pathway, be it immediate or otherwise. This is made clear by the graph theoretical depiction of the DTF results for Example 4.3 (Figure 4.10b), which shows the existence of the influence from $x_1(n)$ onto all other series while it is not influenced by them. As such, PDC portrays immediate direct connectivity, whereas DTF portrays directional signal *reachability* in graph theoretical terms (Harary, 1994), denoting active indirect interaction.

Finally, another important difference between PDC and DTF is their normalization; the former is normalized with respect to the source time series while the latter is normalized with respect to the target time series, thanks to normalization properties similar to that of Equation 4.10.

4.4.3   PDC AND GEWEKE'S MEASURE

Another means of assessing connectivity is based on John Geweke's pioneering work in the multivariate case (Geweke, 1984) and essentially amounts to a two-step procedure whereby a chosen time series is selected, say $x_1(n)$ together with its conjectured causing partner, say $x_2(n)$. The first step consists in fitting a VAR model M_2^r to the restricted series time vector $\mathbf{x}_2^r(n) = [x_1(n)\, x_3(n), \ldots, x_K(n)]^T$ excluding $x_2(n)$. M_2^r's model is combined with $\mathbf{H}(\mathbf{f}) = \bar{\mathbf{A}}^{-1}(f)$ vector moving average representation of the full joint vector time series (the second step) to produce a hybrid model relating $x_2(n)$ and the residuals of the restricted model. Geweke's measure consists of comparing the spectral power of the $x_1^*(n)$ alone to that which can be attributed to $x_2(n)$ in the joint model, that is from $\mathbf{x}_2^*(n) = [x_1^*(n)\, x_2(n)\, x_3^*(n), \ldots, x_K^*(n)]^T$. A modern version of this approach has appeared in Ding et al. (2006) and is slightly modified in Guo et al. (2008), whose aim was to investigate the effect of instantaneous confounders. The reader should be forewarned that the latter paper contains inaccurate PDC estimates that were employed for comparison purposes. See de Brito et al. (2010) and the routine **guo2008_linear.m** in the accompanying CD.

In many ways, Geweke's measure is a hybrid between PDC and DTF as the latter is also based on power ratio comparisons while the former deals with residual time series, even though DTF may be given an information theoretic interpretation

(Takahashi et al., 2010, and Chapter 5). In fact, Geweke's measure has recently been argued as interpretable as a multiple coherence between time series (Chicharro, 2011).

4.4.4 Renormalized PDC

Schelter et al. (2009) suggested a variant of PDC consisting of its numerator $|\bar{A}_{ij}(\lambda)|^2$ normalized with respect to the target structure rather than the source as in PDC. More information about renormalized PDC is available in Chapter 9.

4.4.5 Direct DTF

The product of DTF's absolute value-squared numerator by partial coherence (4.6) was suggested as a means of singling out that part of the effect measured by the signal reachability implied by DTF to only that part taking place directly (Korzeniewska et al., 2003). It is interesting to note that most of the literature does not use Equation 4.6, whose computation is straightforward, but employs the classic definition summarized in Bendat and Piersol (1986), which is applicable to cross-spectral matrices computed via nonparametric estimators, with the necessary consequences to estimator variability.

4.5 DISCUSSION AND CONCLUSIONS

This chapter covered the motivation and some of the properties underlying partial directed coherence, especially in its generalized form, which is scale invariant. Other alternative measures, especially DTF/Directed Coherence, have also been briefly discussed. In fact, in the general case, DTF and PDC describe different properties. The latter pinpoints immediate directed links, whereas nonzero DTF, while stemming from the same VAR time-series model, highlights the presence of signal pathways even if the signal must travel through other structures before reaching the target.

It is not possible to overstress the central role played by model quality, which makes model diagnostics essential. Many factors influence model quality—the correct choice of model order via methods such as Akaike's Information criterion (Akaike, 1974) for starters, the presence of data outliers and the issue of signal stationarity. Ultimately, it is model residue whiteness that needs to be met to ensure the reliability around PDC estimates.

Currently, precise asymptotic statistical behavior is known only for PDC* both in terms of lack of interaction null hypothesis decision thresholds and PDC confidence intervals as discussed in Chapter 7 (see also de Brito et al. 2010; Baccalá et al. 2013). This knowledge makes it attractive as currently all other measures require the use of resampling techniques for statistical appraisal of the sort described in Baccalá et al. (2006).

---

* Some considerations regarding DTF null hypothesis properties by Eichler (2006) are an exception.

Since tests of Granger causality look for evidence associated with $a_{ij}(r) = 0$, a prerequisite for PDC nullity, one can safely claim that PDC is G-causality's frequency domain representation.

REFERENCES

Akaike, H. 1974. A new look at statistical model identification. *IEEE Trans. Autom. Control* *19*: 716–723.

Andrews, D. F. and A. M. Herzberg. 1985. *Data: A Collection of Problems from Many Fields for the Student and Research Worker*. New York: Springer.

Baccalá, L. A., K. Sameshima, G. Ballester et al. 1998. Studying the interaction between brain structures via directed coherence and Granger causality. *Appl. Sig. Process. 5*: 40–48.

Baccalá, L. A. 2001. On the efficient computation of partial coherence from multivariate autoregressive models. In N. Callaos, D. Rosario, and B. Sanches (Eds.), *World Multiconference on Systemics, Cybernetics and Informatics*, Vol. 6, Orlando, 10–14.

Baccalá, L. A. and K. Sameshima. 2001a. Partial directed coherence: A new concept in neural structure determination. *Biol. Cybern. 84*: 463–474.

Baccalá, L. A. and K. Sameshima. 2001b. Partial directed coherence: Some estimation issues. In *World Congress on Neuroinformatics*, 546–553.

Baccalá, L. A. and K. Sameshima. 2001c. Overcoming the limitations of correlation analysis for many simultaneously processed neural structures. *Prog. Brain Res. 130*: 33–47.

Baccalá, L. A., D. Y. Takahashi, and K. Sameshima. 2006. Computer intensive testing for the influence between time series. In *Handbook of Time Series Analysis*, R. Schelter, M. Winterhalter, and J. Timmer (Eds.), 411–436. Weinheim: Wiley-VCH.

Baccalá, L. A., D. Y. Takahashi, and K. Sameshima. 2007. Generalized partial directed coherence. In *15th International Conference on Digital Signal Processing* (DSP2007), 163–166, Cardiff, Wales, UK.

Baccalá, L. A., C. S. N. de Brito, D. Y. Takahashi et al. 2013. Unified asymptotic theory for all partial directed coherence forms. *Philos. Trans. R. Soc. A.* 371: 1–13.

Bendat, J. S. and A. G. Piersol. 1986. *Random Data: Analysis and Measurement Procedures* (2nd ed.). New York: John Wiley.

Chicharro, D. 2011. On the spectral formulation of Granger causality. *Biol. Cybern. 105*: 331–347.

de Brito, C., L. A. Baccalá, D. Y. Takahashi et al. 2010. Asymptotic behavior of generalized partial directed coherence. In *Engineering in Medicine and Biology Society (EMBC), 2010 Annual International Conference of the IEEE*, 1718–1721.

Ding, M. Z. and Y. H. Chen and S. L. Bressler. 2006. Granger causality: Basic theory and application to neuroscience. In *Handbook of Time Series Analysis*, R. Schelter, M. Winterhalter, and J. Timmer (Eds.), 411–436. Weinheim: Wiley-VCH.

Eichenbaum, H. B. and J. L. Davis (Eds.). 1998. *Neuronal Ensembles: Strategies for Recording and Decoding*. New York: John Wiley.

Eichler, M. 2006. On the evaluation of information flow in multivariate systems by the directed transfer function. *Biol. Cybern. 94*: 469–482.

Gersch, W. 1970a. Epileptic focus location: Spectral analysis method. *Science 169*: 701–702.

Gersch, W. 1970b. Spectral analysis of EEG by autoregressive decomposition of time series. *Math. Biosci. 7*: 205–222.

Gersch, W. 1972. Causality or driving in electrophysiological signal analysis. *Math. Biosci. 14*: 177–196.

Geweke, J., R. Meese, and W. Dent. 1983. Comparative tests of causality in temporal systems. *J. Econom. 21*: 161–194.

Geweke, J. 1984. Measures of conditional linear dependence and feedback between time series. *J. Am. Stat. Assoc. 79*: 907–915.

Granger, C. W. J. 1969. Investigating causal relations by econometric models and cross-spectral methods. *Econometrica 37*: 424–438.

Guo, S., J. Wu, M. Ding et al. 2008. Uncovering interactions in the frequency domain. *PLoS Comput. Biol. 4*, e1000087.

Harary, F. 1994. *Graph Theory*. New York: Perseus Books.

Kamiński, M. and K. J. Blinowska. 1991. A new method of the description of the information flow in brain structures. *Biol. Cybern. 65*: 203–210.

Korzeniewska, A., M. Mańczak, M. Kamiński et al. 2003. Determination of information flow direction between brain structures by a modified directed transfer function method (dDTF). *J. Neurosci. Methods 125*: 195–207.

Lütkepohl, H. 1993. *Introduction to Multiple Time Series Analysis* (2nd ed.). Berlin: Springer.

Lütkepohl, H. 2005. *New Introduction to Multiple Time Series Analysis*. New York: Springer.

Nicolelis, M. (Ed.) 1998. *Methods for Neural Ensemble Recordings*. Boca Raton, FL: CRC Press.

Saito, Y. and H. Harashima. 1981. Tracking of information within multichannel record: Causal analysis in EEG. In *Recent Advances in EEG and EMG Data Processing*, 133–146. Amsterdam: Elsevier.

Schelter, B., J. Timmer, and M. Eichler. 2009. Assessing the strength of directed influences among neural signals using renormalized partial directed coherence. *J. Neurosci. Methods 179*: 121–130.

Schnider, S. M., R. H. Kwong, F. A. Lenz et al. 1989. Detection of feedback in the central nervous system using system identification techniques. *Biol. Cybern. 60*: 203–212.

Takahashi, D. Y., L. A. Baccalá, and K. Sameshima. 2010. Information theoretic interpretation of frequency domain connectivity measures. *Biol. Cybern. 103*: 463–469.

5 Information Partial Directed Coherence

Daniel Y. Takahashi, Luiz A. Baccalá, and Koichi Sameshima

CONTENTS

5.1 INTRODUCTION

The advent of time-domain Granger causality (GC) and directed frequency representations for observed neural signals such as PDC has brought the notions of "connectivity" and "information flow" to the fore. Usage of these notions, however, has at best been vague and largely qualitative with diverse often insufficiently substantiated claims as to how one can precisely define information flow quantitatively.

In this chapter, our central question is how does PDC precisely relate to *information flow*? Even though many researchers have sometimes implicitly and sometimes explicitly alluded to it as representing the extent of information flow, very little has, alas, been precisely and rigorously stated about it until recently.

Recent elegant abstract attempts to examine GC from an information theoretic standpoint have appeared: Massey (1990); Schreiber (2000); Hlaváčková-Schindler et al. (2007); and Amblard and Michel (2011). Their common ground has been the employment of the joint probability densities of the observed random signal variables at each instant of time. In practice, despite their fundamental conceptual character, the latter approaches prove difficult to implement because of the stringent estimation accuracy of the required quantities, especially when the observed time series are short. For this reason, we limit ourselves here to the case when linear models are adequate descriptions, even if only approximate, because they allow objective practical results.

As opposed to Takahashi et al. (2010), who should be consulted for proofs and the more technical aspects, here we break up the ideas into their core ingredients by: briefly reviewing some fundamental ideas surrounding multivariate signal representation (Section 5.2) from a fairly elementary standpoint, and later examining the necessary information theoretical aspects that allow the precise statement of what quantities such as DTF and PDC represent in terms of information flow (Section 5.3). The emphasis is on showing that multivariate processes have naturally associated processes: innovation processes and partialized processes, which are helpful in their description and which lead to precise definitions for information flow. Illustrations are performed in Section 5.4 followed by conclusions in Section 5.5.

5.2 PRELIMINARIES

For definiteness, we denote multivariate processes by boldface letters $\mathbf{x}$ and its components by subscripts x_i ($1 \leq i \leq N$), which for simplicity and without loss of generality we consider in the discrete-time domain, that is

$$x_i = \{\dots, x_i(-m), \dots, x(0), \dots, x_i(m), \dots\} \tag{5.1}$$

describes the multivariate process i-th component samples. The notation $\mathbf{x}(n_-)$ refers to all instants before the current $\mathbf{x}(n)$ observation while $\mathbf{x}(n_+)$ denotes the set of future observations.

Two fundamental core ideas must be understood when representing a given process $\mathbf{x}$:

1. It always has an associated *innovation* process $\mathbf{w}$, and
2. When $N > 1$, it has an equivalent representation through its allied *partialized* process $\boldsymbol{\eta}$.

The full probabilistic description of any given process requires the joint probability density between all its describing variables $x_i(k)$ for all i and k. This is intractable in practice except for the simplest cases so that attention is often focused on considering only the joint probabilities between variable pairs $x_i(k)$ and $x_j(l)$. In this case, the term second order or wide-sense description applies. For the Gaussian case, this turns out to be a complete description via the variable means $\mathbb{E}[x_i(k)]$ plus the covariances between variable pairs (computable from the correlations $\mathbb{E}[x_i(m)x_j(l)]$). A process is second order/wide-sense stationary if the latter expectations are invariant to time shifts. This stationarity is what guarantees the existence of the frequency domain descriptions for the processes and constitutes the basis of quantities such as DTF and PDC. Without loss of generality, we consider only the case $\mathbb{E}[x_i(k)] = 0$. In the following, we also employ the notion of conditional expectation.

5.2.1 INNOVATIONS REPRESENTATION

The innovation process $\mathbf{w}$ associated to $\mathbf{x}$ underlies the notion that for regular processes it is impossible to exactly predict $\mathbf{x}(n)$ from $\mathbf{x}(n_-)$, even if a physically arbitrarily precise model with known past observations is available. The current observations can only be predicted with unavoidable random error, which is represented by its associated innovation process **w**.

Hence, in quite some more generality than is perhaps necessary for the current purposes, the impossibility of prediction based on the past may be written as

$$\mathbf{x}(n) = \mathbf{g}(\mathbf{x}(n_-)) + \mathbf{w}(n) = \hat{\mathbf{x}}(n) + \mathbf{w}(n) \tag{5.2}$$

where process current value $\mathbf{x}(n)$ is predicted from $\hat{\mathbf{x}}(n) = \mathbf{g}(\mathbf{x}(n_-))$ up to the innovation $\mathbf{w}(n)$.

It is convenient to stress that

1. It is impossible to predict $\mathbf{w}(n)$ from its own past $\mathbf{w}(n_-)$ (or from its future $\mathbf{w}(n_+)$, for that matter). Or equivalently, $\mathbf{w}(n)$ is independent from $\mathbf{w}(m)$ for $n \neq m$.
2. $\mathbf{g}(\cdot)$ reduces to the conditional expectation of $\mathbf{x}(n)$ given its past, that is $E\left[\mathbf{x}(n)|\mathbf{x}(n_-)\right]$, if $\mathbf{w}(n)$'s variance is to be minimal, that is if one is seeking maximum predictive power for $\hat{\mathbf{x}}(n)$ based on $\mathbf{x}(n_-)$.
3. It is easy to show that if $\mathbf{x}(\cdot)$ is Gaussian*, $\mathbf{g}(\cdot)$ reduces to a linear function of past observations making the VAR (vector autoregressive) models the ones of choice.
4. Naturally, $\mathbf{w}(n)$ is sure to have an impact on future observations on $\mathbf{x}(n_+)$.

Remark 5.1   Last but surely not the least important, $\mathbf{w}(n)$ is often referred to just as "noise" and it is not uncommon to see it mistaken for the totally distinct notion of "additive noise" or "observation noise" whose consideration constitutes an entirely different class of models:

$$\mathbf{y}(n) = \mathbf{h}(\mathbf{x}(n)) + \mathbf{n}(n) \tag{5.3}$$

What distinguishes $\mathbf{n}(n)$ from $\mathbf{w}(n)$ is that it does not directly influence the evolution of $\mathbf{x}(n)$. In practice, $\mathbf{n}(n)$ is always present; since only $\mathbf{y}$ is available, it is used in lieu of $\mathbf{x}$ producing model estimation biases. Even though filtering, given the adequate prior knowledge, can adequately deal with $\mathbf{n}(n)$, this is not pursued here any further.   □

Therefore, to a first linear approximation, the generality of Equation 5.2 translates into the usual VAR process representation:

$$\begin{bmatrix} x_1(n) \\ \vdots \\ x_N(n) \end{bmatrix} = \sum_{r>0} \mathbf{A}_r \begin{bmatrix} x_1(n-r) \\ \vdots \\ x_N(n-r) \end{bmatrix} + \begin{bmatrix} w_1(n) \\ \vdots \\ w_N(n) \end{bmatrix} \tag{5.4}$$

where the discrete time vector process $\mathbf{x}$ is broken into its subscripted component x_i univariate processes.

---

* Gaussianity implies process linearity, but not the converse.

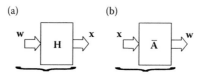

FIGURE 5.1 (a) The use of **H** to generate **x** from **w** is the "moving average" representation of the process (**H, w**) whereas (b) the inverse (prediction error) filter $\bar{\mathbf{A}}$ producing **w** from **x** is its "autoregressive" representation ($\bar{\mathbf{A}}, \mathbf{x}$). Note that $\bar{\mathbf{H}} = \bar{\mathbf{A}}^{-1}$. Representations (a) and (b) are equivalent.

Statistical process stationarity, that is probability density invariance to time shifts, occurs under suitable conditions on $\mathbf{A}_r$ and leads to the usual frequency domain representations of Chapter 4, that is, using

$$\bar{A}_{ij}(\omega) = \begin{cases} 1 - \displaystyle\sum_{r=1}^{p} a_{ij}(r)e^{-\mathbf{j}\omega r}, & \text{if } i = j \\ -\displaystyle\sum_{r=1}^{p} a_{ij}(r)e^{-\mathbf{j}\omega r}, & \text{otherwise} \end{cases} \tag{5.5}$$

with $-\pi < \omega \le \pi^*$ and where $a_{ij}(r)$ is the corresponding $\mathbf{A}_r$ matrix element.

For known $\mathbf{A}_r$ matrices and **x**, one may always compute the innovations process **w**. Conversely, one may generate **x** by directly filtering its innovations **w**. This leads to the equivalent mathematical representations (**H, w**), and (**A, x**) as embodied in the diagrams of Figures 5.1a and 5.1b. These filter representations are the conceptual basis for the developments in Chapter 4 and are respectively associated with DTF and PDC.

It is possible to use the frequency domain representation (5.5) associated with model (5.4) to produce **x**'s spectral representation in terms of its spectral density matrix $\mathbf{S_x}(\omega)$:

$$\begin{bmatrix} S_{x_1 x_1}(\omega) & S_{x_1 x_2}(\omega) & \cdots & \cdots & S_{x_1 x_N}(\omega) \\ \vdots & \ddots & \vdots & S_{x_i x_j}(\omega) & \vdots \\ \vdots & \cdots & S_{x_k x_k}(\omega) & \cdots & \vdots \\ \vdots & \vdots & \vdots & \ddots & \vdots \\ S_{x_N x_1}(\omega) & S_{x_N x_2}(\omega) & \cdots & \cdots & S_{x_N x_N}(\omega) \end{bmatrix} \tag{5.6}$$

which can be straightforwardly written as (Kay, 1988)

$$\mathbf{S_x}(\omega) = \bar{\mathbf{A}}^{-1}(\omega)\mathbf{\Sigma_w}\bar{\mathbf{A}}^{-H}(\omega) = \mathbf{H}(\omega)\mathbf{\Sigma_w}\mathbf{H}^H(\omega) \tag{5.7}$$

where $\bar{\mathbf{A}}(\omega)$ is the matrix collecting the (5.5) quantities and ^{H} the standard Hermitian transpose operation while $\mathbf{\Sigma_w}$ denotes the innovations covariance matrix.

---

* In this chapter, frequencies are normalized to rad/s rather than Hz.

One may as well use the entries in Equation 5.6 to compute the coherences $C_{x_i x_j}(\omega)$ (Bendat and Piersol, 1986) between process components in terms of cross- $(S_{x_i x_j}(\omega))$ and auto- $(S_{x_i x_i}(\omega), S_{x_j x_j}(\omega))$ spectra as

$$C_{x_i x_j}(\omega) = \frac{S_{x_i x_j}(\omega)}{\sqrt{S_{x_i x_i}(\omega) S_{x_j x_j}(\omega)}} \tag{5.8}$$

which plays a fundamental role expressing mutual information between the two component subprocesses x_i and x_j (Section 5.2.3).

5.2.2 PARTIALIZATION

The innovations representation exists even if a process is univariate ($N = 1$). Partialization, by contrast, needs at least $N = 2$ component processes.

The partialized process is constructed by collecting partialized components η_i associated with each component process x_i via

$$\eta_i = x_i - \mathbb{E}[x_i | \mathbf{x}^i] \tag{5.9}$$

where $\mathbf{x}^i$ stands for the set of $\mathbf{x}$'s components excluding the x_i component. In other words, η_i is that intrinsic part of x_i that cannot be predicted using the other components of $\mathbf{x}^i$ of $\mathbf{x}$.

One can show that η_i can be obtained by Wiener filtering $\mathbf{x}$ (Haykin, 2001):

$$\eta_i = x_i - \mathbf{g}_i^T \mathbf{x}^i \tag{5.10}$$

where the row filter operator vector is given by

$$\mathbf{g}_i^T(\omega) = \mathbf{s}_{x_i \mathbf{x}^i}(\omega) \mathbf{S}_{\mathbf{x}^i \mathbf{x}^i}^{-1}(\omega) \tag{5.11}$$

where $\mathbf{S}_{\mathbf{x}^i \mathbf{x}^i}(\omega)$ is the spectral density matrix filter operator associated with $\mathbf{x}^i$ and

$$\mathbf{s}_{x_i \mathbf{x}^i}(\omega) = [S_{x_i x_1}(\omega) \ \ldots \ S_{x_i x_{i-1}}(\omega) \ S_{x_i x_{i+1}}(\omega) \ \ldots \ S_{x_i x_N}(\omega)] \tag{5.12}$$

is a row operator vector associated with cross-spectra between x_i and the other $\mathbf{x}^i$ time series.

Consequently, one can write the spectral density of η_i as

$$S_{\eta_i \eta_i}(\omega) = S_{x_i x_i}(\omega) - \mathbf{s}_{x_i \mathbf{x}^i}(\omega) \mathbf{S}_{\mathbf{x}^i \mathbf{x}^i}^{-1}(\omega) \mathbf{s}_{\mathbf{x}^i x_i}(\omega) \tag{5.13}$$

It is important to keep in mind that the Wiener solution represented by Equation 5.11 produces the minimum mean-squared error to the prediction of x_i when the other process components are used. This means that η_i represents the dynamics of x_i after stripping it from the information of all other observed processes. In other words, η_i is *orthogonal* to $\mathbf{x}^i$ and this constitutes a key factor in exposing the directional information associated with PDC.

Furthermore, one may show that

$$S_{\eta_i \eta_i}^{-1}(\omega) = \mathbf{a}_i^H(\omega) \mathbf{\Sigma}_{\mathbf{w}}^{-1} \mathbf{a}_i(\omega) \tag{5.14}$$

where $\bar{\mathbf{a}}_i(\omega)$ is the i-th column of $\bar{\mathbf{A}}(\omega)$.

Note that it is also possible to partialize $\mathbf{w}(n)$ as well by computing

$$\zeta_i = w_i - \mathbb{E}[w_i | \mathbf{w}^i] \tag{5.15}$$

using

$$\zeta_i = w_i - \boldsymbol{\sigma}_{w_i \mathbf{w}^i}^T \mathbf{\Sigma}_{\mathbf{w}^i}^{-1} \mathbf{w}^i \tag{5.16}$$

whose variance is given by

$$\sigma_{\zeta_i \zeta_i} = \sigma_{w_i w_i} - \boldsymbol{\sigma}_{w_i \mathbf{w}^i}^T \mathbf{\Sigma}_{\mathbf{w}^i}^{-1} \boldsymbol{\sigma}_{w_i \mathbf{w}^i} \tag{5.17}$$

where $\boldsymbol{\sigma}_{w_i \mathbf{w}^i}$ is the vector collecting the cross covariances between w_i and the other components $\mathbf{w}^i$ of $\mathbf{w}$ whose intrinsic covariances are collected in the appropriate $\mathbf{\Sigma}_{\mathbf{w}^i}^{-1}$ taken from the $\mathbf{\Sigma}_{\mathbf{w}}$ covariance matrix.

Here, partialization decomposes the innovation w_i into ζ_i, which represents the innovation variability that may come from no other $\mathbf{w}^i$ innovation component. As such, ζ_i is orthogonal to $\mathbf{w}^i$, that is, unless $i = j$, the covariance is null:

$$COV(\zeta_i, w_j) = 0 \tag{5.18}$$

The time argument n in the $\mathbf{w}$ and $\boldsymbol{\zeta}$ components above was omitted thanks to their presumed stationarity.

5.2.3 INFORMATION THEORY

The idea of information is closely attached to the probability of a random event. Mechanisms for generating random events (that is, random variables) are characterized by their *entropy*, that is the average information they can generate. The comparison of how much information is common to two random mechanisms (variables) X and Y is made via their *mutual information*, which is given by

$$I(X, Y) = \sum_{x,y} \mathbb{P}(x, y) \log \frac{\mathbb{P}(x, y)}{\mathbb{P}(x)\mathbb{P}(y)} \tag{5.19}$$

where $\mathbb{P}(x)$ and $\mathbb{P}(y)$ are the probabilities (marginals) associated respectively with standalone X and Y, whereas $\mathbb{P}(x, y)$ describes their joint random mechanisms, such that $\mathbb{P}(x, y) = \mathbb{P}(x)\mathbb{P}(y)$ when they are independent, implying the nullity of Equation 5.19, which translates to lack of relatedness.

Time series can be placed within the same framework as they can be viewed as a succession of (possibly related) random variables in time as in Equation 5.1, which is rigorously described by the joint probability density $\mathbb{P}(x_i)$ of its observations. Comparison between the contents of two time series can be achieved through their mutual entropy rates:

$$
\begin{aligned}
\text{MIR}(x, y) = \lim_{m \to \infty} \frac{1}{m+1} \\
\times \mathbb{E}\left[\log \frac{d\mathbb{P}(x_i(1), \ldots, x_i(m), x_j(1), \ldots, x_j(m))}{d\mathbb{P}(x_i(1), \ldots, x_i(m))d\mathbb{P}(x_j(1), \ldots, x_j(m))}\right]
\end{aligned}
\tag{5.20}
$$

which generalizes $I(X, Y)$ to reflect the interaction between time series and which settles down to a constant value when the series are jointly stationary.

In their fundamental paper, Gelfand and Yaglom (1959) showed that Equation 5.20 can be written in terms of the coherence $C_{x_j x_j}(\omega)$ between the time series:

$$
\text{MIR}(x_i, x_j) = -\frac{1}{4\pi} \int_{-\pi}^{\pi} \log(1 - |C_{x_i x_j}(\omega)|^2)\, d\omega
\tag{5.21}
$$

for jointly Gaussian stationary time series.

Note how the integrand $-\log(1 - |C_{x_i x_j}(\omega)|^2) \geq 0$ for all ω. This allows this quantity to be interpreted as the frequency domain increments that must be added up to produce the total mutual information rate $\text{MIR}(x_i, x_j)$.

5.3   MAIN RESULTS

From the standpoint of inferring interaction directionality between x_i and x_j, Equations 5.20 and 5.21 are unhelpful given their symmetry, that is,

$$
\text{MIR}(x_i, x_j) = \text{MIR}(x_j, x_i)
\tag{5.22}
$$

something that follows from the equality

$$
|C_{x_i x_j}(\omega)|^2 = |C_{x_j x_i}(\omega)|^2
\tag{5.23}
$$

On the other hand, as discussed in Section 5.2, time-series pairs can be broken down into more fundamental processes: its associated innovations w_i and its partialized versions η_i.

Our core result was to show that suitably generalized forms of PDC and DTF can be written in terms of the mutual information between the underlying innovation/partialized processes that alternatively describe $\mathbf{x}$ (Takahashi et al., 2010) thereby providing a clear interpretation to the latter measures in information terms.

One can show (Takahashi et al., 2010) that the squared coherence between the innovation associated with the i-th series, w_i, and the partialized version of the j-th series, η_j, is given by

$$
|C_{w_i \eta_j}(\omega)|^2 = |\iota \pi_{ij}(\omega)|^2
\tag{5.24}
$$

where

$$\iota\pi_{ij}(\omega) = \sigma_{w_i w_i}^{-1/2} \frac{\bar{A}_{ij}(\omega)}{\sqrt{\bar{a}_j^H(\omega)\Sigma_w^{-1}\bar{a}_j(\omega)}} \tag{5.25}$$

which simplifies to the originally defined PDC when Σ_w equals the identity matrix (see Chapter 4). Generalized PDC (gPDC) from Baccalá et al. (2007) is likewise obtained if only the main diagonal of Σ_w is employed.

Therefore, one can write

$$\text{MIR}(w_i, \eta_j) = -\frac{1}{4\pi} \int_{-\pi}^{\pi} \log(1 - |\iota\pi_{ij}(\omega)|^2) \, d\omega \tag{5.26}$$

which relates the innovations associated with the i-th series and the stripped down version of x_j represented by its partialized process η_j. What is solely x_j's and does not come from any other series has its communality with the w_i innovations described by Equation 5.24 or equivalently in information units by Equation 5.26.

The interaction in the opposite direction is represented by

$$|C_{w_j \eta_i}(\omega)|^2 = |\iota\pi_{ji}(\omega)|^2 \tag{5.27}$$

which comes from the decomposition of different component processes and thus describes the interaction in the reverse direction.

Similarly, one may show (Takahashi et al., 2010) that the coherence

$$|C_{x_i \zeta_j}(\omega)|^2 = |\iota DTF_{ij}(\omega)|^2 \tag{5.28}$$

where

$$\iota DTF_{ij}(\omega) = \sigma_{\zeta_j \zeta_j}^{1/2} \frac{H_{ij}(\omega)}{\sqrt{\mathbf{h}_j^H(\omega)\Sigma_w \mathbf{h}_j(\omega)}}, \tag{5.29}$$

where $H_{ij}(\omega)$ are the entries in the $\mathbf{H}(\omega)$ matrix in Equation 5.7 and $\mathbf{h}_j(\omega)$ is a vector of its rows. Equation 5.29 reduces to the usual form of the DTF (Chapters 2 and 4) when $\Sigma_w = \mathbf{I}$.

Naturally,

$$\text{MIR}(x_i, \zeta_j) = -\frac{1}{4\pi} \int_{-\pi}^{\pi} \log(1 - |\iota DTF_{ij}(\omega)|^2) \, d\omega \tag{5.30}$$

measures how much of the stripped down (partialized) innovations ζ_j has in common with the observed time series x_i. Much as in the PDC case, swapping i and j involves relating different underlying processes and thus represents the interaction in the reverse direction.

5.4 ILLUSTRATIONS

Consider the following simple model

$$\begin{bmatrix} x_1(n) \\ x_2(n) \end{bmatrix} = \begin{bmatrix} 0 & 0 \\ \alpha & 0 \end{bmatrix} \begin{bmatrix} x_1(n-1) \\ x_2(n-1) \end{bmatrix} + \begin{bmatrix} w_1(n) \\ w_2(n) \end{bmatrix} \tag{5.31}$$

where $\Sigma_w = I$ where $\iota\pi_{12}(\omega) = 0$ and

$$\iota\pi_{21}(\omega) = \frac{-\alpha e^{-j\omega}}{\sqrt{1+\alpha^2}}$$

can be obtained by direct computation using Equation 5.25 or may be alternatively computed directly through $C_{w_1\eta_2}(\omega)$ using the fact that

$$s_{21}(\omega)S_{11}^{-1}(\omega) = \alpha e^{-j\omega}$$

that is $\eta_2(n) = x_2(n) - \alpha x_1(n-1) = w_2(n)$, which implies that $C_{w_1\eta_2}(\omega) = 0 = \iota\pi_{12}$ since $COV(w_1, w_2) = 0$.

To compute $C_{w_2\eta_1}(\omega)$, one must use the spectral density matrix of $[x_1 \ x_2]^T$ given by

$$\begin{bmatrix} S_{x_1x_1}(\omega) & S_{x_1x_2}(\omega) \\ S_{x_2x_1}(\omega) & S_{x_2x_2}(\omega) \end{bmatrix} = \begin{bmatrix} 1 & \alpha e^{j\omega} \\ \alpha e^{-j\omega} & 1+\alpha^2 \end{bmatrix} \tag{5.32}$$

which leads to the Wiener filter

$$G_1(\omega) = s_{12}(\omega)S_{22}^{-1}(\omega) = \frac{\alpha}{1+\alpha^2}e^{j\omega}$$

so that

$$\eta_1(n) = x_1(n) - \frac{\alpha}{1+\alpha^2}x_2(n+1) \tag{5.33}$$

but which leads to

$$\eta_1(n) = w_1(n)\frac{1}{1+\alpha^2} - w_2(n+1)\frac{\alpha}{1+\alpha^2} \tag{5.34}$$

However, since $x_1(n) = w_1(n)$ and $x_2(n) = \alpha w_1(n-1) + w_2(n)$ implies

$$S_{w_2\eta_1}(\omega) = \frac{-\alpha e^{-j\omega}}{1+\alpha^2} \tag{5.35}$$

and

$$S_{\eta_1}(\omega) = \frac{1}{1+\alpha^2}$$

$$S_{w_2}(\omega) = 1$$

showing that

$$C_{w_2\eta_1}(\omega) = \frac{-\alpha e^{-j\omega}}{\sqrt{1+\alpha^2}}$$

confirming that $\iota\pi_{21}(\omega) = C_{w_2\eta_1}(\omega)$ via direct computation.

Enlarging the Equation 5.31 with a third observed variable

$$x_3(n) = \beta x_2(n-1) + w_3(n) \tag{5.36}$$

where the $w_3(n)$ innovation is independent, the zero mean unit variance Gaussian implies that the signal x_1 has an indirect path to x_3 via x_2 and still no direct means of reaching x_3.

Under the hypothesis of uncorrelated unit variance innovations, it is easy to compute this model's connectivity as

$$\iota\pi_{21}(\omega) = \frac{-\alpha e^{-j\omega}}{\sqrt{1+\alpha^2}},$$

$$\iota\pi_{32}(\omega) = \frac{-\beta e^{-j\omega}}{\sqrt{1+\beta^2}},$$

$$\iota\pi_{31}(\omega) = 0, \tag{5.37}$$

confirming through Equation 5.37 that no direct connection exists.

More cumbersome calculations are needed when the innovations are correlated and of different variance. In this case, the relative behavior of gPDC to ιPDC is provided in Figure 5.2 for this example's $i = 1$ to $j = 2$ connection, showing that no mandatory magnitude relationship between gPDC to ιPDC necessarily exists.

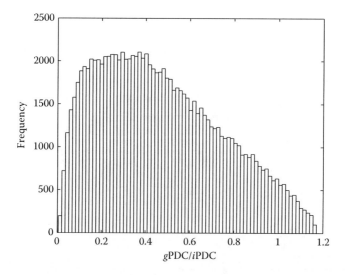

FIGURE 5.2 Ratio of gPDC to ιPDC from $i = 1$ to $j = 2$ for 100,000 randomly chosen innovation covariance matrices is shown. The histogram was obtained for $\omega = \pi/4$.

5.5   COMMENTS AND CONCLUSIONS

In this chapter, we have summarized how different process description methods (via $\mathbf{H}$ or $\bar{\mathbf{A}}$) filters lead to means of computing the relationship between innovation components and partialized representations via their coherences, which constitute generalizations of DTF and PDC and can thus be used to express the mutual information rates between process portions.

The logarithms in the integrands of Equations 5.26 and 5.30 represent information units of measurement for the directed relationships represented by PDC and DTF, respectively.

The chief characteristic of both Equations 5.26 and 5.30 is to adequately consider the prediction error covariance matrix $\mathbf{\Sigma}_w$ to weigh the relationships into correctly scaled information measurements, which matter only when the innovations are either strongly correlated or when their variances differ considerably in size.

A final interesting point that surfaces from the examination of the proof (see Takahashi et al., 2010) of Equation 5.24 regards PDC's numerator

$$\sigma_{w_i w_i}^{-1/2} \bar{A}_{ij}(\omega) \qquad (5.38)$$

which is readily amenable to interpretation as a transfer function relating w_i to η_j. A reweighted version of Equation 7.20 has been used recently by Schelter et al. (2009) to describe connectivity under a different normalization to portray connectivity (see Chapter 9).

At this point, it is perhaps fair to recall that PDC, gPDC, ιPDC, and Equation 7.20 allow the problem of connectivity detection under appropriate criteria to be addressed, and it is treated in more detail in Chapter 7. Among the latter measures, it is only ιPDC that properly accounts for size effects in gauging connection strength, given its ability to express information flow in rigorous fashion at the expense of the normalization properties (4.2) and (4.10).

REFERENCES

Amblard, P.-O. and O. Michel. 2011. On directed information theory and Granger causality graphs. *J. Comput. Neurosci. 30*: 7–16.

Bendat, J. S. and A. G. Piersol. 1986. *Random Data: Analysis and Measurement Procedures* (2nd ed.). New York: John Wiley.

Baccalá, L. A., D. Y. Takahashi, and K. Sameshima. 2007. Generalized partial directed coherence. In *15th International Conference on Digital Signal Processing*, 163–166.

Gelfand, I. M. and A. M. Yaglom. 1959. Calculation of amount of information about a random function contained in another such function. *Am. Math. Soc. Transl. Ser. 2*: 3–52.

Haykin, S. 2001. *Adaptive Filter Theory*, (4th ed). Upper Saddle River: Prentice-Hall.

Hlaváčková-Schindler, K., M. Paluš, M. Vejmelka et al. 2007. Causality detection based on information-theoretic approaches in time series analysis. *Phys. Rep. 441*: 1–46.

Kay, S. M. 1988. *Modern Spectral Estimation*. Englewood Cliffs: Prentice-Hall.

Massey, J. 1990. Causality, feedback and directed information. In *Proc. 1990 Int. Symp. on Info. Th. Its Appl. (ISITA'90)*, pp. 303–305, Hawaii.

Schelter, B., J. Timmer, and M. Eichler. 2009. Assessing the strength of directed influences among neural signals using renormalized partial directed coherence. *J. Neurosci. Methods* *179*: 121–130.

Schreiber, T. 2000. Measuring information transfer. *Phys. Rev. Lett. 85*: 461–464.

Takahashi, D. Y., L. A. Baccalá, and K. Sameshima. 2010. Information theoretic interpretation of frequency domain connectivity measures. *Biol. Cybern. 103*: 463–469.

6 Assessing Connectivity in the Presence of Instantaneous Causality

Luca Faes

CONTENTS

6.1 INTRODUCTION

This chapter is devoted to a discussion of the impact of *instantaneous causality* on the computation of frequency domain connectivity measures. Instantaneous causality (IC) refers to interactions between two observed time series which occur within the same time lag. While IC is hard to interpret in real-world physical systems because causal interactions take time to occur in these systems, it becomes a clear concept when time-series models are used to describe physical interactions in practice. In fact, instantaneous effects show up in practical time-series modeling anytime the time resolution of the measurements is lower than the time scale of the lagged causal influences occurring among the processes underlying the measured time series. This situation is likely encountered in experimental time-series analysis and, in particular, cannot be ignored in the study of brain connectivity.

The linear parametric model implemented in the definition of the main frequency domain connectivity measures and in the numerous following applications (see e.g., Baccalá et al., 1998; Kaminski et al., 2001; Astolfi et al., 2006; Baccalá and Sameshima, 2001; Winterhalder et al., 2005) is a strictly causal Multivariate autoregressive (MVAR) model; strict causality means that only lagged effects among the observed time series are modeled, while instantaneous effects—which are descriptive of IC—are not described by any model coefficient. Neglecting instantaneous effects in MVAR models implies that any existing zero-lag correlation among the time series is translated into a correlation among the model innovations (Lütkepohl, 1993). Note that two of the most popular frequency domain causality measures, that is, the PDC and the DTF (Baccalá and Sameshima, 2001; Kaminski et al., 2001), as well as their generalizations including the innovation variances (Baccalá et al. 1998; 2007), explicitly forsake cross-covariances among the MVAR model innovations. Thus, when these measures are used, IC is omitted from the frequency domain evaluation of causality. This choice was made because instantaneous effects are not directly related to the concept of Granger causality, which is indeed based on an exclusive consideration of lagged effects (Granger, 1969). However, it is known that instantaneous correlations may severely affect the classical strictly causal MVAR representation of multiple time series (Lütkepohl, 1993), and this may have an adverse impact on the estimation of connectivity from the model coefficients. Specifically, the interpretation of the MVAR coefficients as causal effects, which is implicit in the development of any frequency domain causality measure, presupposes that the model innovations are independent of each other. If this is not the case, the causal interpretation may not be justified because there is some structure in the observed data that has not been modeled through the MVAR coefficients.

To overcome the above limitations, we introduce here an extended MVAR model, which combines both lagged and instantaneous effects as a new framework for the evaluation of connectivity in the frequency domain. The novelty with respect to traditional strictly causal modeling is in the fact that in the extended representation instantaneous effects are explicitly described as coefficients of the model, so that independence of the model innovations is achieved by construction. We show how the spectral representation of the extended model may be exploited to quantify in the frequency domain a variety of connectivity definitions in multivariate processes: coupling (i.e., nondirectional interactions), lagged causality according to the Granger definitions (1969, 1980), and extended forms of causality which include instantaneous effects in addition to lagged ones. We categorize known and novel frequency domain connectivity measures, derived from strictly causal and extended MVAR models, according to the different time-domain definitions of connectivity which they underlie. Then, we compare the spectral profiles of these measures computed on theoretical examples of multivariate closed-loop processes with and without instantaneous interactions. Finally, we discuss aspects of model identification, paying particular attention to the estimation of the coefficients of the extended MVAR model, which constitutes a challenging problem that can be solved only by imposition of additional constraints on the distribution of the data under analysis.

6.2   TIME-DOMAIN CONNECTIVITY DEFINITIONS

Let us consider the zero mean stationary vector process $\mathbf{x} = [x_1 \cdots x_K]^T$. Given the two scalar components x_i and x_j, the general concept of *connectivity* can be particularized to the study of *causality* or *coupling* between x_i and x_j, which investigate respectively the directional or nondirectional properties of the considered pairwise interaction. In the following, we provide definitions of causality and coupling for a general vector process, and then we particularize these definitions for MVAR processes.

6.2.1   MULTIVARIATE CLOSED-LOOP PROCESSES

To introduce the notation with respect to the kth scalar process composing the vector process $\mathbf{x}$, denote $x_k(n)$ as the current value of x_k, $X_k(n) = \{x_k(n-1), \dots, x_k(n-p)\}$ as the set of the p past values of x_k, and as $\bar{X}_k(n) = \{x_k(n), X_k(n)\} = \{x_k(n), x_k(n-1), \dots, x_k(n-p)\}$ as the set of the p past values of x_k enlarged with the current value. Then, $\mathbf{x}$ is a multivariate closed-loop process of order p if, for each $k = 1, \dots, K$,

$$x_k(n) = f_k\left(X_k(n), \bar{\mathbf{X}}^k(n)\right) + w_k(n), \tag{6.1}$$

where $\bar{\mathbf{X}}^k(n) = \left[\bar{X}_{l_1}(n) \cdots \bar{X}_{l_{K-1}}(n)\right]^T$, $\{l_1, \dots, l_{K-1}\} = \{1, \dots, K\}\backslash\{k\}$, and w_k is a zero mean stationary innovation process describing the error in the representation. The multivariate closed-loop process is defined by the set of functions f_k in Equation 6.1, which relate the current value of each considered scalar process to the set of its own past values and the present and past values of all other processes; this set of values, which we denote as $\mathbf{U}^k(n) = \{X_k(n), \bar{\mathbf{X}}^k(n)\}$, represents the most complete information set that can be used to describe $x_k(n)$. In order for the definition (6.1) to be unambiguous, the functions f_k must fully explain the dynamical interactions within the vector process $\mathbf{x}$; this condition corresponds to making the assumption that the scalar innovation processes w_k ($k = 1, \dots, K$) are mutually independent, both of each other and over time.

The concepts of causality and coupling may be defined in terms of prediction improvement, that is, reduction of the variance of the prediction error w_k resulting from modifying the information set on which the prediction model is conditioned. Accordingly, with the aim of supporting the interpretation of the frequency domain connectivity measures defined in Section 6.3, the following definitions are provided for the two scalar processes x_i and x_j, $i, j = 1, \dots, K$, $i \neq j$:

Instantaneous causality (IC) from x_j to x_i, $x_j \overset{\circ}{\to} x_i$ exists if the prediction of $x_i(n)$ based on $\mathbf{U}^i(n)$ yields a lower prediction error than the prediction based on $\mathbf{U}^i(n)\backslash x_j(n)$.

Lagged direct causality (LDC) from x_j to x_i, $x_j \to x_i$ exists if the prediction of $x_i(n)$ based on $\mathbf{U}^i(n)$ yields a lower prediction error than the prediction based on $\mathbf{U}^i(n)\backslash X_j(n)$.

Extended direct causality (EDC) from x_j to x_i, $x_j \overset{*}{\to} x_i$ exists if the prediction of $x_i(n)$ based on $\mathbf{U}^i(n)$ yields a lower prediction based on $\mathbf{U}^i(n)\backslash\bar{X}_j(n)$.

Lagged total causality (LTC) from x_j to x_i, $x_j \Rightarrow x_i$ exists if a cascade of L Lagged
   direct causality (LDC) relations occurs such that $x_{k_0} \to \cdots \to x_{k_L}$ with $k_0 =
   j, k_L = i, \{k_1, \ldots, k_{L-1}\} \in \{1, \ldots, K\} \backslash \{i, j\}, 1 \le L \le K - 1$.
Extended total causality (ETC) from x_j to x_i, $x_j \overset{*}{\Rightarrow} x_i$ exists if a cascade of L
   Extended direct causality (EDC) relations occurs such that $x_{k_0} \overset{*}{\to} \cdots \overset{*}{\to} x_{k_L}$
   with $k_0 = j, k_L = i, \{k_1, \ldots, k_{L-1}\} \in \{1, \ldots, K\} \backslash \{i, j\}, 1 \le L \le K - 1$.
Direct coupling (DCoup) between x_j and x_i, $x_i \leftrightarrow x_j$ exists if $x_i \overset{*}{\to} x_j$ and/or $x_j \overset{*}{\to} x_i$.
Coupling (Coup) between x_j and x_i, $x_i \Leftrightarrow x_j$ exists if $x_i \overset{*}{\Rightarrow} x_j$ and/or $x_j \overset{*}{\Rightarrow} x_i$ for at
   least one $k \in \{1, \ldots, K\}$.

The classical IC and LDC, which reflect well-defined causality concepts originally introduced by Granger (1969, 1980). In particular, Instantaneous causality (IC)
refers to the exclusive consideration of instantaneous effects (i.e., from $x_j(n)$ to $x_i(n)$),
while LDC refers to the exclusive consideration of lagged effects (i.e., from $X_j(n)$ to
$x_i(n)$) in determining the prediction improvements. The definition of EDC is a generalization which explicitly combines both instantaneous and lagged effects according
to recently emerged ideas (Faes and Nollo, 2010a,b; Hyvarinen et al., 2010; Faes
et al., 2013). Furthermore, the definitions of total causality, LTC, and ETC, generalize
LDC and EDC, respectively, by considering indirect effects, that is, effects mediated
by one or more other processes in the multivariate closed loop, in addition to the
direct effects between the two considered scalar processes. Finally, the two coupling
definitions generalize the corresponding causality definitions by accounting for both
the forward and backward effects (either instantaneous and lagged) between the two
considered processes.

In addition to the definitions provided above, we state the following coupling
definitions, which are referred to as spurious because they concern a mathematical
formalism rather than an intuitive property of two interacting processes: *spurious
direct coupling* between x_j and x_i exists if $x_i \overset{*}{\to} x_k$ and $x_j \overset{*}{\to} x_k$ for at least one
$k \in \{1, \ldots, K\}, k \ne i, k \ne j$; *spurious coupling* between x_j and x_i exists if $x_k \overset{*}{\Rightarrow} x_i$
and $x_k \overset{*}{\Rightarrow} x_j$ for at least one $k \in \{1, \ldots, K\}, k \ne i, k \ne j$. These definitions suggest
that two processes can also be interpreted as directly coupled when they both directly
cause a third common process and, as coupled, when they are both caused by a third
common process, respectively. The definitions of spurious Direct coupling (DCoup)
and Coupling (Coup) are introduced here to provide a formalism that will explain
a confounding property of the two common frequency-domain coupling measures
reviewed in Section 6.3 (i.e., the coherence and the partial coherence).

6.2.2 MULTIVARIATE AUTOREGRESSIVE PROCESSES

In the linear signal processing framework, the considered vector process **x** is commonly represented by an MVAR model of order p:

$$\mathbf{x}(n) = \sum_{l=1}^{p} \mathbf{A}(l)\mathbf{x}(n-l) + \mathbf{u}(n), \qquad (6.2)$$

where $\mathbf{A}(l) = \left\{a_{ij}(l)\right\}_{i,j=1}^{K}$ is the coefficient matrix at lag l, and $\mathbf{u} = [u_1 \cdots u_K]^T$ is a vector of K zero mean innovation processes with positive-definite zero lag covariance matrix $\mathbf{\Sigma}_u \equiv \mathbf{\Sigma} = \mathbb{E}[\mathbf{u}(n)\mathbf{u}^T(n)] = \{\sigma_{ij}\}_{i,j=1}^{K}$. The MVAR representation (6.2) is a particularization of Equation 6.1 in which each function f_k is a linear first-order polynomial. This representation is *strictly causal*, in the sense that only lagged effects among the processes are modeled, while instantaneous effects are not described by any model coefficient because $\mathbf{A}(l)$ is defined only for positive lags ($l = 0$ is not considered in Equation 6.2). Hence, zero-lag correlations among the processes x_i and x_j, when present, remain as correlations among the model innovations u_i and u_j, so that the innovation covariance $\mathbf{\Sigma}$ is generally not diagonal ($\sigma_{ij} = \sigma_{ji} \neq 0$) (Lütkepohl, 1993). Note that this particular situation violates the assumption of mutual independence of the innovations stated in Section 6.2.1 for vector closed-loop processes.

The representation in Equation 6.2 allows investigation of properties of the joint description of the processes composing $\mathbf{x}$ from the analysis of the MVAR coefficients. For instance, the definitions of LDC and Lagged total causality (LTC) provided in Section 6.2.1 for a general closed-loop process may be formulated for a strictly causal MVAR process as follows: $x_j \rightarrow x_i$ if $a_{ij}(l) \neq 0$ for at least one $l \in \{1, \ldots, p\}$; $x_j \Rightarrow x_i$ if $a_{k_m k_{m-1}}(l_m) \neq 0, m = 1, \ldots, L$, with $k_0 = j, k_L = i$, for some $l_1, \ldots, l_L \in \{1, \ldots, p\}$. Note that IC cannot be tested in a similar way, because instantaneous effects are not modeled in Equation 6.2. Therefore, strictly causal MVAR models do not represent IC and, consequently, EDC and Extended total causality (ETC), in terms of the model coefficients. Nevertheless, with the strictly causal representation, one can test the lack of IC by showing that the innovation covariance $\mathbf{\Sigma}$ is diagonal. Moreover, the information about zero lag correlations contained in the off-diagonal elements of $\mathbf{\Sigma}$ may be exploited for testing DCoup and Coup, as we will see in Section 6.3.1.

As an alternative to using the strictly causal model (6.2), the vector process $\mathbf{x}$ can be described by including instantaneous effects in the interactions allowed by the model. This is achieved by considering the extended MVAR model (Faes and Nollo, 2010a,b; Faes et al., 2012; Hyvarinen et al., 2010):

$$\mathbf{x}(n) = \sum_{l=0}^{p} \mathbf{B}(l)\mathbf{x}(n-l) + \mathbf{w}(n), \tag{6.3}$$

where $\mathbf{B}(l) = \{b_{ij}(l)\}_{i,j=1}^{K}$, and $\mathbf{w} = [w_1 \cdots w_K]^T$ is a vector of K zero mean uncorrelated innovations with diagonal zero lag covariance matrix $\mathbf{\Sigma}_w \equiv \mathbf{\Lambda} = \mathbb{E}[\mathbf{w}(n)\mathbf{w}^T(n)] = diag\{\lambda_{ii}\}_{i=1}^{K}$. The difference with respect to strictly causal MVAR modeling as in Equation 6.2 is that now the lag variable l takes the value 0 as well, which brings instantaneous effects from $x_j(n)$ to $x_i(n)$ into the model in the form of the coefficients $b_{ij}(0)$ of the matrix $\mathbf{B}(0)$. In the extended MVAR model, absence of correlation among the noise inputs w_k, that is, diagonality of $\mathbf{\Lambda}$, is guaranteed by the presence of the instantaneous effects.

In contrast to the strictly causal MVAR model which may describe only lagged interactions, the extended MVAR representation (6.3) allows the detection of all types of connectivity defined in Section 6.2.1 in terms of the elements $b_{ij}(l)$ of the coefficient matrices $\mathbf{B}(l)$. Specifically, the basic definitions of IC, LDC, and EDC are stated as follows: $x_j \overset{\circ}{\to} x_i$, $x_j \to x_i$, and $x_j \overset{*}{\to} x_i$ occur respectively when $b_{ij}(0) \neq 0$, $b_{ij}(l) \neq 0$ for at least one $l \in \{1,\ldots,p\}$, and $b_{ij}(l) \neq 0$ for at least one $l \in \{0,1,\ldots,p\}$. Then, the other definitions follow through generalizations to cascades of direct effects and/or consideration of both forward and backward effects: for instance, LTC $x_j \Rightarrow x_i$ and ETC $x_j \overset{*}{\Rightarrow} x_i$ occur when $b_{k_m k_{m-1}}(l_m) \neq 0$, with $k_0 = j$, $k_L = i$, for a set of lags $l_1,\ldots,l_L$ with values in $\{1,\ldots,p\}$ and with values in $\{0,1,\ldots,p\}$, respectively. DCoup $x_i \leftrightarrow x_j$ occurs when $b_{ki}(l_1) \neq 0$ and $b_{kj}(l_2) \neq 0$ for at least one $k \in \{1,\ldots,K\}$ and at least two lags $l_1, l_2 \in \{0,1,\ldots,p\}$; if $k = i$ or $k = j$, the relation is spurious. Coup $x_i \leftrightarrow x_j$ occurs when $b_{k_m k_{m-1}}(l_m) \neq 0$, both with $k_0 = k, k_L = i$, and $k_0 = k, k_L = j$, for at least one $k \in \{1,\ldots,K\}$ and at least two sets of lags $l_1,\ldots,l_L$ with values in $\{0,1,\ldots,p\}$; if $k = i$ or $k = j$, the relation is spurious.

The relation between the strictly causal representation and the extended representation can be established moving the term $\mathbf{B}(0)\mathbf{x}(n)$ from the right-hand side to the left-hand side of Equation 6.3 and then left-multiplying both sides by the matrix $\mathbf{L}^{-1} = [\mathbf{I} - \mathbf{B}(0)]$, where $\mathbf{I}$ is the $K \times K$ identity matrix. Then, the comparison with Equation 6.2 yields:

$$\mathbf{A}(l) = \mathbf{L}\mathbf{B}(l) = [\mathbf{I} - \mathbf{B}(0)]^{-1}\mathbf{B}(l) \tag{6.4a}$$

$$\mathbf{u}(n) = \mathbf{L}\mathbf{w}(n), \quad \mathbf{\Sigma} = \mathbf{L}\mathbf{\Lambda}\mathbf{L}^T \tag{6.4b}$$

Note that in the absence of IC (i.e., when $\mathbf{B}(0) = 0$) the extended model (6.3) reduces to the strictly causal model (6.2), as in this case we have $\mathbf{L} = \mathbf{I}, \mathbf{A}(l) = \mathbf{B}(l)$, $\mathbf{u}(n) = \mathbf{w}(n)$, $\mathbf{\Sigma} = \mathbf{\Lambda}$. In this case, EDC reduces to LDC and ETC reduces to LTC. By contrast, the presence of instantaneous effects modeled by $\mathbf{B}(0) \neq 0$ makes all coefficients $\mathbf{B}(l)$ different from $\mathbf{A}(l)$. Hence, although LDC and LTC are defined in complete analogy with the two representations (i.e., considering only lagged effects), different LDC and LTC patterns may be found depending on whether the strictly causal model or the extended model is used to describe the considered vector process.

6.3 FREQUENCY DOMAIN CONNECTIVITY MEASURES

In the following, we derive connectivity measures that reflect and quantify in the frequency domain the time-domain definitions provided in Section 6.2. First, the known correlation and partial correlation time-domain analyses are transposed into the frequency domain for describing the concepts of coupling and direct coupling, respectively. Second, the parametric representation of the vector process is exploited to decompose the derived spectral measures of (direct) coupling into measures of (direct) causality.

As to the first step, the time-domain interactions within the vector closed-loop process $\mathbf{x}$ may be characterized by means of the correlation matrix $\mathbf{R}(l) = \mathbb{E}[\mathbf{x}(n)\mathbf{x}^T(n-l)]$ and of its inverse $\mathbf{R}^{-1}(l)$, whose elements are combined to compute the correlation coefficient and partial correlation coefficient, respectively (Whittaker, 1990). The correlation and partial correlation quantify respectively the overall linear interdependence between two scalar processes, and their linear interdependence after removing the effects of all remaining processes through a procedure denoted as partialization (Brillinger et al., 1976). As such, these quantities reflect Coup and DCoup in the time domain. Their frequency domain counterpart is obtained through the traditional spectral analysis of vector processes, on the one hand (Kay, 1988), and through a corresponding dual analysis performed in the inverse spectral domain on the other (Dahlhaus, 2000). Specifically, the $K \times K$ spectral density matrix $\mathbf{S}(\omega) = \{S_{ij}(\omega)\}_{i,j=1}^{K}$ is defined as the Fourier transform (FT) of the correlation matrix $\mathbf{R}(l)$, while the inverse spectral matrix $\mathbf{P}(\omega) = \mathbf{S}^{-1}(\omega) = \{P_{ij}(\omega)\}_{i,j=1}^{K}$ results as the Fourier's transform (FT) of the partial correlation matrix $\mathbf{R}^{-1}(l)$ ($\omega \in [-\pi, \pi)$ is the frequency measured in radians). The two spectral matrices are to define coherence (Coh) and partial coherence (PCoh) between x_i and x_j as follows (Eichler et al., 2003):

$$\Gamma_{ij}(\omega) = \frac{S_{ij}(\omega)}{\sqrt{S_{ii}(\omega)S_{jj}(\omega)}}, \tag{6.5a}$$

$$\Pi_{ij}(\omega) = -\frac{P_{ij}(\omega)}{\sqrt{P_{ii}(\omega)P_{jj}(\omega)}}. \tag{6.5b}$$

Note that in view of their being frequency domain analogues of correlation and partial correlation, Coh and PCoh can be interpreted as spectral measures of coupling and direct coupling. In the next subsections, we show that when a multivariate closed loop process is reduced to a linear parametric MVAR process, Coh and PCoh reflect the Coup and DCoup definitions given in Section 6.2. Moreover, in this case the MVAR coefficients may be used to decompose the measures of coupling defined in Equation 6.5 into corresponding measures of causality; this can be done by exploiting either the strictly causal or the extended MVAR representations, resulting in different frequency domain measures.

6.3.1 MEASURES BASED ON STRICTLY CAUSAL MVAR MODELS

The spectral representation of the strictly causal MVAR process is obtained by taking the FT of Equation 6.2 to yield $\mathbf{X}(\omega) = \mathbf{A}(\omega)\mathbf{X}(\omega) + \mathbf{U}(\omega)$, where $\mathbf{X}(\omega)$ and $\mathbf{U}(\omega)$ are the FTs of $\mathbf{x}(n)$ and $\mathbf{u}(n)$, and $\mathbf{A}(\omega) = \sum_{k=1}^{p} \mathbf{A}(k)e^{-j\omega k}$ is the $K \times K$ frequency domain coefficient matrix (here $\mathbf{j} = \sqrt{-1}$). If one wants to evidence the transfer function from $\mathbf{u}(n)$ to $\mathbf{x}(n)$, the spectral representation becomes $\mathbf{X}(\omega) = \mathbf{H}(\omega)\mathbf{U}(\omega)$, where $\mathbf{H}(\omega) = [\mathbf{I} - \mathbf{A}(\omega)]^{-1} = \bar{\mathbf{A}}^{-1}(\omega)$ is the $K \times K$ frequency domain transfer matrix.

The two spectral representations above, evidencing the coefficient matrix $\bar{\mathbf{A}}(\omega) = \{\bar{A}_{ij}(\omega)\}_{i,j=1}^{K}$ and the transfer matrix $\mathbf{H}(\omega) = \{H_{ij}(\omega)\}_{i,j=1}^{K}$, can be exploited to express the spectral density matrix and its inverse, according to a well-known spectral factorization theorem (Gevers and Anderson, 1981), in terms of the frequency domain coefficient and transfer matrices as

$$\mathbf{S}(\omega) = \mathbf{H}(\omega)\mathbf{\Sigma}\mathbf{H}^{H}(\omega) \tag{6.6a}$$

$$\mathbf{P}(\omega) = \bar{\mathbf{A}}^{H}(\omega)\mathbf{\Sigma}^{-1}\bar{\mathbf{A}}(\omega), \tag{6.6b}$$

where ^{H} stands for the Hermitian transpose. The elements of $\mathbf{S}(\omega)$ and $\mathbf{P}(\omega)$ can be represented in compact form as

$$S_{ij}(\omega) = \mathbf{h}_i(\omega)\mathbf{\Sigma}\mathbf{h}_j^H(\omega) \tag{6.7a}$$

$$P_{ij}(\omega) = \bar{\mathbf{a}}_i^H(\omega)\mathbf{\Sigma}^{-1}\bar{\mathbf{a}}_j(\omega), \tag{6.7b}$$

where $\mathbf{h}_i(\omega)$ is the ith row of $\mathbf{H}(\omega)$, $\mathbf{h}_i(\omega) = [H_{i1}(\omega)\cdots H_{iK}(\omega)]$, and $\bar{\mathbf{a}}_j(\omega)$ is the jth column of $\bar{\mathbf{A}}(\omega)$, $\bar{\mathbf{a}}_j(\omega) = \left[\bar{A}_{1j}(\omega)\cdots\bar{A}_{Kj}(\omega)\right]^T$.

Under the assumption of absence of instantaneous effects, which results in a diagonal form for the zero lag covariance matrix $\mathbf{\Sigma}$ and for its inverse, Equations 6.7a and 6.7b can be further expressed as

$$S_{ij}(\omega) = \sum_{k=1}^{K} \sigma_{kk}H_{ik}(\omega)H_{jk}^*(\omega) \tag{6.8a}$$

$$P_{ij}(\omega) = \sum_{k=1}^{K} \sigma_{kk}^{-1}\bar{A}_{ki}^*(\omega)\bar{A}_{kj}(\omega). \tag{6.8b}$$

The usefulness of the factorizations in Equation 6.8 is in the fact that they allow the decomposition of the frequency domain measures of coupling and direct coupling previously defined in terms eliciting the directional information from one process to another. Indeed, substituting Equations 6.7 and 6.8 into Equation 6.5, the Coh and PCoh between x_i and x_j can be factored as

$$\Gamma_{ij}(\omega) = \sum_{k=1}^{K} \iota\gamma_{ik}(\omega)\iota\gamma_{jk}^*(\omega), \tag{6.9a}$$

$$\Pi_{ij}(\omega) = -\sum_{k=1}^{K} \iota\pi_{kj}(\omega)\iota\pi_{ki}^*(\omega), \tag{6.9b}$$

where Equations 6.9a and 6.9b contain the information directed transfer function iDTF and the information partial directed coherence iPDC, which are defined from x_j to x_i as (Takahashi et al., 2010):

$$\iota\gamma_{ij}(\omega) = \frac{\sigma_{jj}^{1/2}H_{ij}(\omega)}{\sqrt{S_{ii}(\omega)}}, \tag{6.10a}$$

$$\iota\pi_{ij}(\omega) = \frac{\sigma_{ii}^{-1/2}\bar{A}_{ij}(\omega)}{\sqrt{P_{jj}(\omega)}}. \tag{6.10b}$$

We remark that, although the iDTF and iPDC can always be computed from the coefficients of a strictly causal MVAR process, they can be interpreted as factors in the decomposition of Coh and PCoh, according to Equation 6.9, only in the absence of instantaneous effects; this is because the decompositions in Equation 6.8 hold only when Σ and Σ^{-1} are diagonal matrices. Under the assumption of the absence of IC, the squared modulus of iDTF and of iPDC assume the meaningful interpretation of the relative amount of power of the output process which is due to the input process, and of the relative amount of inverse power of the input process which is sent to the output process, respectively:

$$S_{ii}(\omega) = \sum_{k=1}^{K} S_{i|k}(\omega), \; S_{i|k}(\omega) = |\iota\gamma_{ik}(\omega)|^2 S_{ii}(\omega), \tag{6.11a}$$

$$P_{jj}(\omega) = \sum_{k=1}^{K} P_{j\to k}(\omega), \; P_{j\to k}(\omega) = |\iota\pi_{kj}(\omega)|^2 P_{jj}(\omega). \tag{6.11b}$$

The properties stated in Equation 6.11 correspond to the normalization properties of iDTF and iPDC, $\sum_{k=1}^{K} |\iota\gamma_{ik}(\omega)|^2 = 1$ and $\sum_{k=1}^{K} |\iota\pi_{kj}(\omega)|^2 = 1$, which therefore hold only when zero lag correlations between the observed processes may be reasonably excluded.

Note also that, in the absence of IC, Equations 6.10a and 6.10b reduce to the formulations of the directed coherence, which we denote here as generalized directed transfer function (gDTF), and of the generalized partial directed coherence (gPDC) as defined, respectively, by Baccalá et al. (1998, 2007):

$$g\gamma_{ij}(\omega) = \frac{\sigma_{jj}^{1/2} H_{ij}(\omega)}{\sqrt{\sum_{k=1}^{K} \sigma_{kk} |H_{ik}(\omega)|^2}}, \tag{6.12a}$$

$$g\pi_{ij}(\omega) = \frac{\sigma_{ii}^{-1/2} \bar{A}_{ij}(\omega)}{\sqrt{\sum_{k=1}^{K} \sigma_{kk}^{-1} |\bar{A}_{kj}(\omega)|^2}}. \tag{6.12b}$$

Moreover, with the additional assumption of equal variance for all the scalar innovations, the two measures further reduce to the original Directed transfer function (DTF), $\gamma_{ij}(\omega)$, and Partial directed coherence (PDC), $\pi_{ij}(\omega)$, as defined respectively by Kaminski et al. (2001) and Baccalá and Sameshima (2001). These measures always satisfy the normalization conditions, that is, $\sum_{k=1}^{K} |g\gamma_{ik}(\omega)|^2 = 1$, $\sum_{k=1}^{K} |\gamma_{ik}(\omega)|^2 = 1$, $\sum_{k=1}^{K} |g\pi_{kj}(\omega)|^2 = 1$, and $\sum_{k=1}^{K} |\pi_{kj}(\omega)|^2 = 1$; however, we remark that a meaningful interpretation like that given in Equation 6.11 holds only for the gDTF and the gPDC, and, again, only under the assumption of absence of IC.

The correspondence of the frequency domain connectivity measures presented up to now with the time-domain connectivity definitions provided in Section 6.2 is summarized in Table 6.1. The iPDC is a measure of LDC, because the numerator of Equation 6.10b, when $i \neq j$, is nonzero only when $a_{ij}(l) \neq 0$ for some lag l (i.e., when $x_j \to x_i$) and is uniformly zero when $a_{ij}(l) = 0$ for each l; the same holds also for the

TABLE 6.1
Connectivity Definitions and Frequency Domain Measures

Connectivity Definition	Symbol	scMVAR	eMVAR
		iPDC, $\iota\pi_{ij}$	lPDC, $\bar{\chi}_{ij}$
Lagged direct causality (LDC)	$x_j \rightarrow x_i$	gPDC, $g\pi_{ij}$	nPDC, $n\bar{\chi}_{ij}$
		PDC, π_{ij}	
		iDTF, $\iota\gamma_{ij}$	lDTF, $\tilde{\xi}_{ij}$
Lagged total causality (LTC)	$x_j \Rightarrow x_i$	gDTF, $g\gamma_{ij}$	nDTF, $n\tilde{\xi}_{ij}$
		DTF, γ_{ij}	
Extended direct causality (EDC)	$x_j \overset{*}{\rightarrow} x_i$	—	ePDC, χ_{ij}
Extended total causality (ETC)	$x_j \overset{*}{\Rightarrow} x_i$	—	eDTF, ξ_{ij}
Direct coupling (DCoup)	$x_j \leftrightarrow x_i$	PCoh, Π_{ij}	
Coupling (Coup)	$x_j \Leftrightarrow x_i$	Coh, Γ_{ij}	

Note: scMVAR, strictly causal MVAR representation; eMVAR, extended MVAR representation.

gPDC and the PDC. The iDTF is a measure of LTC, as one can show that expanding the transfer function $H_{ij}(\omega)$ as a geometric series results in a sum of terms where each one is related to one of the (direct or indirect) transfer paths connecting x_j to x_i (Eichler, 2006). Therefore, the numerator of Equation 6.10a is nonzero whenever at least one path connecting x_j to x_i is significant, that is, when $x_j \Rightarrow x_i$; the same holds also for the generalized directed transfer function (gDTF) and the DTF.

As for Coh and PCoh, we note from Equations 6.9a and 6.9b that $\Gamma_{ij}(\omega) \neq 0$ when both $\iota\gamma_{ik}(\omega) \neq 0$ and $\iota\gamma_{jk}(\omega) \neq 0$, and that $\Pi_{ij}(\omega) \neq 0$ when both $\iota\pi_{ki}(\omega) \neq 0$ and $\iota\pi_{kj}(\omega) \neq 0$; moreover, note that $S_{ij}(\omega) \neq 0$ and $P_{ij}(\omega) \neq 0$, so that $\Gamma_{ij}(\omega) \neq 0$ and $\Pi_{ij}(\omega) \neq 0$, also when the $i - j$ elements of Σ and Σ^{-1} are nonzero (see Equation 6.7). This suggests that Coh and PCoh reflect, respectively, coupling and direct coupling relations in accordance with a frequency domain representation of the definitions given in Section 6.2. However, Equations 6.9a and 6.9b also explain the rationale of introducing a mathematical formalism to define spurious coupling and spurious direct coupling (Faes et al., 2012). In fact, the fulfillment of $\Gamma_{ij}(\omega) \neq 0$ or $\Pi_{ij}(\omega) \neq 0$ at a given frequency ω is not a sufficient condition for the existence of Coup and DCoup at that frequency, because the observed relation can also be spurious.

6.3.2 Measures Based on Extended MVAR Models

The spectral representation of the extended MVAR process is obtained taking the FT of Equation 6.3 to yield $\mathbf{X}(\omega) = \mathbf{B}(\omega)\mathbf{X}(\omega) + \mathbf{W}(\omega)$, where $\mathbf{B}(\omega) = \mathbf{B}(0) + \sum_{l=1}^{p} \mathbf{B}(l)e^{-\mathbf{j}\omega l}$. The input–output spectral representation becomes $\mathbf{X}(\omega) = \mathbf{G}(\omega)\mathbf{W}(\omega)$, where $\mathbf{G}(\omega) = [\mathbf{I} - \mathbf{B}(\omega)]^{-1} = \bar{\mathbf{B}}^{-1}(\omega)$. Given these representations evidencing the frequency domain coefficient matrix $\bar{\mathbf{B}}(\omega) = \left\{\bar{B}_{ij}(\omega)\right\}_{i,j=1}^{K}$ and transfer matrix $\mathbf{G}(\omega) = \left\{G_{ij}(\omega)\right\}_{i,j=1}^{K}$, the spectral matrix and its inverse can be factored for the extended MVAR representation as

$$\mathbf{S}(\omega) = \mathbf{G}(\omega)\mathbf{\Lambda}\mathbf{G}^{H}(\omega), \tag{6.13a}$$

$$\mathbf{P}(\omega) = \bar{\mathbf{B}}^{H}(\omega)\mathbf{\Lambda}^{-1}\bar{\mathbf{B}}(\omega). \tag{6.13b}$$

By means of some matrix algebra involving the spectral representations of Equations 6.2 and 6.3, as well as Equation 6.4b, it is easy to show that the spectral matrix (and its inverse as well) resulting in Equation 6.13 are exactly the same as those obtained in Equation 6.6. This demonstrates the equivalence of the spectral representation for strictly causal and extended MVAR processes. Consequently, the concepts of coupling and direct coupling are also equivalent for the two process representations, since Coh and PCoh computed as in Equation 6.5 depend exclusively on the elements of $\mathbf{S}(\omega)$ and $\mathbf{P}(\omega)$. For this reason, common Coh and PCoh measures are indicated in Table 6.1 for strictly causal and extended MVAR models.

A substantial difference between the two representations arises when coupling relations are decomposed to infer causality. In fact, for the extended MVAR representation, the elements of $\mathbf{S}(\omega)$ and $\mathbf{P}(\omega)$ can be represented as

$$S_{ij}(\omega) = \mathbf{g}_i(\omega)\mathbf{\Lambda}\mathbf{g}_j^{H}(\omega) = \sum_{k=1}^{K} \lambda_{kk} G_{ik}(\omega)G_{jk}^{*}(\omega), \tag{6.14a}$$

$$P_{ij}(\omega) = \bar{\mathbf{b}}_i^{H}(\omega)\mathbf{\Lambda}^{-1}\bar{\mathbf{b}}_j(\omega) = \sum_{k=1}^{K} \lambda_{kk}^{-1} \bar{B}_{ki}^{*}(\omega)\bar{B}_{kj}(\omega), \tag{6.14b}$$

where $\mathbf{g}_i(\omega) = [G_{i1}(\omega)\cdots G_{iK}(\omega)]$ and $\bar{\mathbf{b}}_j(\omega) = \left[\bar{B}_{1j}(\omega)\cdots\bar{B}_{Kj}(\omega)\right]^{T}$, without requiring assumptions on the form of the zero lag innovation covariance matrix. The factorizations in Equation 6.14 are indeed valid still in the presence of instantaneous interactions among the observed processes, because the extended MVAR representation leads to diagonal input covariance matrices $\mathbf{\Lambda}$ and $\mathbf{\Lambda}^{-1}$ by construction. Therefore, using the extended MVAR representation it is always possible to decompose the Coherence (Coh) and Partial coherence (PCoh) between x_i and x_j as

$$\Gamma_{ij}(\omega) = \sum_{k=1}^{K} \xi_{ik}(\omega)\xi_{jk}^{*}(\omega), \tag{6.15a}$$

$$\Pi_{ij}(\omega) = -\sum_{k=1}^{K} \chi_{kj}(\omega)\chi_{ki}^{*}(\omega), \tag{6.15b}$$

where the factors in the decompositions (6.15a) and (6.15b) contain the extended directed transfer function (eDTF) and the extended partial directed coherence (*e*PDC), which are defined from x_j to x_i as (Faes and Nollo, 2010b; Faes et al., 2013):

$$\xi_{ij}(\omega) = \frac{\lambda_{jj}^{1/2} G_{ij}(\omega)}{\sqrt{S_{ii}(\omega)}}, \tag{6.16a}$$

$$\chi_{ij}(\omega) = \frac{\lambda_{ii}^{-1/2} \bar{B}_{ij}(\omega)}{\sqrt{P_{jj}(\omega)}}. \tag{6.16b}$$

Thanks to the inclusion of instantaneous effects in the MVAR model, the following decompositions hold both in the presence and the absence of IC among the scalar processes composing $\mathbf{x}$:

$$S_{ii}(\omega) = \sum_{k=1}^{K} S_{i|k}(\omega), \, S_{i|k}(\omega) = |\xi_{ik}(\omega)|^2 S_{ii}(\omega), \tag{6.17a}$$

$$P_{jj}(\omega) = \sum_{k=1}^{K} P_{j \to k}(\omega), \, P_{j \to k}(\omega) = |\chi_{kj}(\omega)|^2 P_{jj}(\omega). \tag{6.17b}$$

Therefore, *e*DTF and *e*PDC can always be interpreted, respectively, as the relative amount of power of the output process which is due to the input process, and as the relative amount of inverse power of the input process which is sent to the output process. Note that the information quantified by these extended measures is both lagged ($l > 0$) and instantaneous ($l = 0$), because it is computed in the frequency domain from the function $\mathbf{B}(\omega)$ which incorporates both $\mathbf{B}(0)$ and $\mathbf{B}(l)$ with $l > 0$. As a consequence, *e*DTF and *e*PDC reflect in the frequency domain the definitions of ETC and EDC, respectively (see Table 6.1).

To perform exclusive analysis of the lagged causal effects in the presence of zero-lag interactions, one has to exclude the coefficients related to the instantaneous effects from the desired spectral causality measure. This can be done by setting the frequency domain coefficient and transfer matrices that contain the lagged effects only:

$$\tilde{\mathbf{B}}(\omega) = \bar{\mathbf{B}}(\omega) + \mathbf{B}(0) = \mathbf{I} - \sum_{l=1}^{p} \mathbf{B}(l)e^{-\mathbf{j}\omega l}, \tag{6.18a}$$

$$\tilde{\mathbf{G}}(\omega) = \tilde{\mathbf{B}}^{-1}(\omega), \tag{6.18b}$$

and then using these matrices in analogy with Equations 6.10 and 6.16 to define the so-called lagged directed transfer function (*l*DTF) and lagged partial directed coherence (*l*PDC):

$$\tilde{\xi}_{ij}(\omega) = \frac{\lambda_{jj}^{1/2} \tilde{G}_{ij}(\omega)}{\sqrt{S_{ii}(\omega)}}, \tag{6.19a}$$

$$\tilde{\chi}_{ij}(\omega) = \frac{\lambda_{ii}^{-1/2} \tilde{B}_{ij}(\omega)}{\sqrt{P_{jj}(\omega)}}, \tag{6.19b}$$

or in analogy with Equation 6.12 to define the so-called normalized lagged directed transfer function (nDTF) and normalized lagged partial directed coherence (nPDC) (Faes and Nollo, 2010b; Faes et al., 2013):

$$n\tilde{\xi}_{ij}(\omega) = \frac{\lambda_{jj}^{1/2}\tilde{G}_{ij}(\omega)}{\sqrt{\sum_{k=1}^{K}\lambda_{kk}|\tilde{G}_{ik}(\omega)|^2}}, \qquad (6.20a)$$

$$n\tilde{\chi}_{ij}(\omega) = \frac{\lambda_{ii}^{-1/2}\tilde{B}_{ij}(\omega)}{\sqrt{\sum_{k=1}^{K}\lambda_{kk}^{-1}|\tilde{B}_{kj}(\omega)|^2}}. \qquad (6.20b)$$

Since the measures in Equations 6.19 and 6.20 are derived using exclusively the time-domain matrices of lagged effects, we have lDTF and nDTF measure LTC, while lPDC and nPDC measure LDC, in the frequency domain (see Table 6.1). Note that the lDTF and lPDC, such as the corresponding measures derived from a strictly causal model (i.e., iDTF and iPDC), do not satisfy normalization conditions as in Equation 6.11 or in Equation 6.17 unless IC is absent. The nDTF and nPDC, like the corresponding strictly causal measures gDTF and gPDC, are always normalized but still lack a meaningful interpretation as in Equation 6.11 or Equation 6.17 because their denominator differs from a diagonal element of $\mathbf{S}(\omega)$ or $\mathbf{P}(\omega)$, unless IC is absent. Therefore, a full meaningful interpretation of all frequency domain measures of lagged causality can be achieved only in the absence of instantaneous effects; in such a condition, all measures of LDC defined in Table 6.1 simplify to a single gPDC (and to the original PDC when all innovation variances coincide), while all measures of LTC simplify to a single gDTF (and to the original DTF when all innovation variances coincide).

Finally, we stress that, even though they reflect in the frequency domain the same causality definition, lPDC and nPDC may indicate different LDC patterns than iPDC, gPDC, and PDC, while lDTF and nDTF may indicate different LTC patterns than iDTF, gDTF, and DTF. This is because the strictly causal and the extended MVAR model representations may produce different lagged causality patterns from their time-lagged coefficients. In particular, we will see in the next section that the erroneous interpretation of lagged effects provided by the strictly causal model in the presence of instantaneous effects may lead to misleading profiles of the various frequency domain measures of lagged causality, while the correct interpretation is retrieved using the analogous measures derived from the extended model.

6.4 THEORETICAL EXAMPLE

To illustrate the various connectivity definitions and frequency domain measures defined above, and compare them in the presence or absence of instantaneous causality, let us consider the exemplary process of order $p = 2$:

$$\begin{cases} x_1(n) = 0.9\sqrt{2}x_1(n-1) - 0.81x_1(n-2) + w_1(n) \\ x_2(n) = 0.5x_1(n-\delta) + x_3(n-1) + w_2(n) \\ x_3(n) = 0.5x_2(n-\delta) + 0.5x_2(n-2) - 0.64x_3(n-2) + w_3(n), \end{cases} \quad (6.21)$$

where the innovations w_k are uncorrelated white noises with variance $\lambda_{kk} = k$, ($k = 1, 2, 3$). The processes x_1 and x_3 present autonomous oscillations, determined by the coefficients of the regression on their own past values, at the frequencies $\omega_1 = \pi/4$ and $\omega_3 = \pi/2$. Then, direct causal effects are set from x_3 to x_2 with lag 1, from x_2 to x_3 with lag 2, as well as from x_1 to x_2 and from x_2 to x_3 with lag δ. In the next two subsections, we consider the conditions $\delta = 1$ and $\delta = 0$, which entail the absence and presence of IC, respectively.

6.4.1 Strictly Causal MVAR Process

When we set $\delta = 1$ in Equation 6.21, all causal effects among the three processes are lagged in time. In this case, IC is absent and the extended causality definitions, EDC and ETC, reduce to the lagged definitions, LDC and LTC. According to Equation 6.21, the imposed LDC relations are $x_1 \rightarrow x_2$, $x_2 \rightarrow x_3$, and $x_3 \rightarrow x_2$, which determine the LTC relations $x_1 \Rightarrow x_2, x_2 \Rightarrow x_3, x_3 \Rightarrow x_2$ (direct effects), and $x_1 \Rightarrow x_3$ (indirect effect). The corresponding Coup and DCoup relations involve all pairs of processes, that is, we have $x_1 \leftrightarrow x_2, x_2 \leftrightarrow x_3, x_1 \leftrightarrow x_3$ (spurious DCoup effect), and $x_1 \nleftrightarrow x_2, x_2 \nleftrightarrow x_3, x_1 \nleftrightarrow x_3$.

The connectivity relations listed above can be easily deducted from the MVAR representations of the process (6.21), which are depicted in Figure 6.1. In this case, the strictly causal and extended MVAR representations are equivalent, because the absence of IC entails $\mathbf{B}(0) = 0$ and thus, according to Equation 6.4, we have $\mathbf{A}(l) = \mathbf{B}(l), \forall l \geq 1$ and $\mathbf{u} = \mathbf{w}$, and $\mathbf{\Sigma} = \mathbf{\Lambda}$ are diagonal matrices. Therefore, the analysis of connectivity of this process without instantaneous effects can be performed alike through strictly causal modeling or through extended modeling.

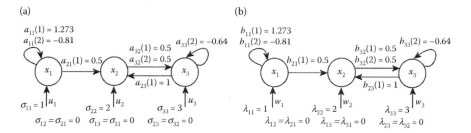

FIGURE 6.1 Strictly causal MVAR representation (a) and extended MVAR representation (b) of the exemplary vector process generated by the equations in Equation 6.21 with parameter $\delta = 1$.

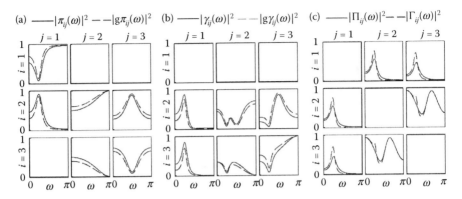

FIGURE 6.2 Diagrams of (a) PDC (solid lines) and gPDC (dashed lines), (b) DTF (solid lines) and gDTF (dashed lines), (c) PCoh (solid lines) and Coh (dashed lines), computed as a squared modulus between each pair of scalar processes x_i and x_j of the illustrative example (6.21) with $\delta = 1$.

Figure 6.2 reports the profiles of the various frequency domain connectivity measures (squared modulus) computed for the theoretical example (6.21) with $\delta = 1$. In this case, with the absence of IC, the measures of extended and lagged causality computed from the extended model reduce to the measures of lagged causality computed from the strictly causal model. For this reason, only the generalized PDC and DTF measures are shown in Figure 6.2, while all other measures (i.e., information partial directed coherence (iPDC), lPDC, nPDC, ePDC, and information directed transfer function (iDTF), lDTF, nDTF, eDTF) overlap with gPDC and gDTF, respectively. As shown in Figure 6.2a, the PDC measures reflect in the frequency domain the presence or absence of direct causal effects, as both PDC and gPDC exhibit nonzero profiles when LDC is set between two processes and are uniformly zero when LDC is not set. In a similar way, the DTF measures reflect total causality effects, as both DTF and gDTF are nonzero only when LTC is set between two processes (Figure 6.2b). Note that in this example the profiles of PDC do not overlap with those of gPDC, and the profiles of DTF do not overlap with those of gDTF, because the three innovation processes in Equation 6.21 have a different variance. Finally, the nonzero profiles of Coh and PCoh in all diagrams of the matrix layout plot of Figure 6.2c indicate the existence of Coup and DCoup between each pair of processes. Note the symmetry of the two functions ($\Gamma_{ij}(\omega) = \Gamma_{ji}^*(\omega)$, $\Pi_{ij}(\omega) = \Pi_{ji}^*(\omega)$), reflecting the fact that they cannot account for the directionality of the considered pairwise interactions.

6.4.2 EXTENDED MVAR PROCESS

Instantaneous effects are induced in the considered exemplary process by setting $\delta = 0$ in Equation 6.21. In this case, the imposed IC and LDC relations are $x_1 \overset{\circ}{\to} x_2$, $x_2 \overset{\circ}{\to} x_3$, and $x_2 \to x_3$, $x_3 \to x_2$, respectively, which determine the EDC relations $x_1 \overset{*}{\to} x_2$, $x_2 \overset{*}{\to} x_3$, and $x_3 \overset{*}{\to} x_2$. The corresponding total causality relations are $x_2 \Rightarrow x_3$, $x_3 \Rightarrow x_2$ (LTC), and $x_1 \overset{*}{\Rightarrow} x_2$, $x_2 \overset{*}{\Rightarrow} x_3$, $x_3 \overset{*}{\Rightarrow} x_2$, $x_1 \overset{*}{\Rightarrow} x_3$ (ETC). The DCoup

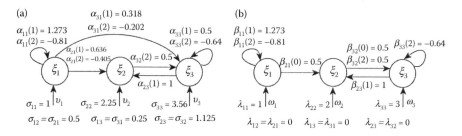

FIGURE 6.3 Strictly causal MVAR representation (a) and extended MVAR representation (b) of the exemplary vector process generated by the equations in Equation 6.21 with parameter $\delta = 0$.

and Coup relations are the same as before, that is, $x_1 \leftrightarrow x_2$, $x_2 \leftrightarrow x_3$, $x_1 \leftrightarrow x_3$, and $x_1 \Leftrightarrow x_2$, $x_2 \Leftrightarrow x_3$, $x_1 \Leftrightarrow x_3$.

In this case with $\delta = 0$, the process (6.21) is suitably described as an extended MVAR process in the form of Equation 6.3, whereby the use of coefficients describing both instantaneous and lagged causal effects allows the whole set of connectivity relations to be represented (see Figure 6.3b). In contrast, when a strictly causal MVAR process in the form of Equation 6.2 is used to describe the same network, the resulting model is that of Figure 6.3a. The strictly causal structure results from the application of Equation 6.4 to the extended structure, leading to different values for the coefficients. As seen in Figure 6.3a, the result is an overestimation of the number of active direct pathways, and a general different estimation of the causality patterns. In particular, the strictly causal model erroneously interprets as lagged the causal effect from x_1 to x_2, which is actually instantaneous, and erroneously detects a spurious direct causality effect from x_1 to x_3, which is actually absent. The misleading connectivity pattern of Figure 6.3a is the result of the impossibility for the strictly causal model (6.2) to describe instantaneous effects: according to Equation 6.4b, in the strictly causal representation, these effects are translated into the input covariance matrix: indeed, not only the innovation variances are different, but also cross-correlations between the input processes arise, as indicated in Figure 6.3a.

The problems of using the strictly causal MVAR representation in the presence of instantaneous effects become evident when one computes the frequency domain connectivity measures. Indeed, the spectral representations closely reflect the time-domain diagrams, but—quite for this reason—only the connectivity pictures derived from the extended model are meaningful, while those derived from the strictly causal model may be strongly misleading. This is demonstrated in Figure 6.4, depicting the frequency domain evaluation of connectivity for the considered theoretical example. As shown in Figures 6.4a and 6.4b, utilization of the strictly causal representation leads to erroneous interpretation of the frequency domain patterns of lagged causality. Indeed, all measures of LDC (i.e., PDC, gPDC, and iPDC) erroneously indicate as lagged not only the instantaneous direct connection from x_1 to x_2, but also the presence of a direct connection from x_1 to x_3 that is actually neither instantaneous nor lagged (Figure 6.4a). All measures of LTC (i.e., DTF, gDTF, and iDTF) provide

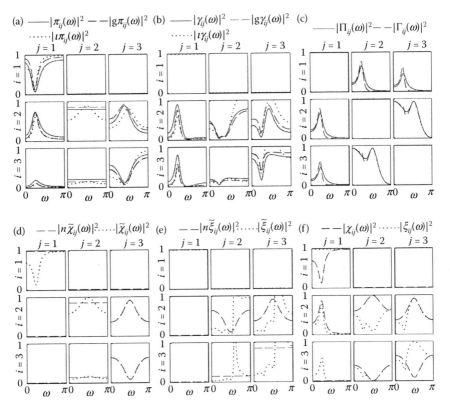

FIGURE 6.4 Diagrams of (a) PDC (solid lines), gPDC (dashed lines), and iPDC (dotted), (b) DTF (solid lines), gDTF (dashed lines), and iDTF (dotted), (c) PCoh (solid lines) and Coh (dashed lines), (d) nPDC (dashed lines) and lPDC (dotted), (e) nDTF (dashed lines) and lDTF (dotted), (f) ePDC (dashed lines) and eDTF (dotted), computed as a squared modulus for each pair of scalar processes x_i and x_j of the illustrative example (6.21) with $\delta = 0$.

similar misinterpretation (Figure 6.4b). In contrast, computation of lagged causality from the coefficients of an extended MVAR model leads to a correct frequency domain interpretation, as documented in Figure 6.4d for the LDC measures (i.e., nPDC and lPDC) and in Figure 6.4e for the LTC measures (i.e., nDTF and lDTF). In this case, we see that only the connections from x_2 to x_3 and from x_3 to x_2 are characterized by nonzero frequency domain causality. Moreover, the extended representation allows a thorough assessment of extended causality to be performed in the frequency domain, as documented in Figure 6.4f where we see that the squared ePDC and eDTF functions exhibit nonzero spectral profiles respectively when EDC and ETC are set between two processes. Finally, in Figure 6.4c, we report the frequency domain profiles of Coup and DCoup measured by the squared Coh and PCoh functions, and remark that—due to the equivalence of Equations 6.6 and 6.13—these profiles perfectly overlap when computed either from the strictly causal or from the extended MVAR representation.

6.5   PRACTICAL ANALYSIS

The practical computation of frequency domain connectivity from a set of K time series of N samples, $x_k(n), k = 1, \ldots, K; n = 1, \ldots, N$, measured from a physical system, is based on considering the series as a finite length realization of the vector process $\mathbf{x} = [x_1 \cdots x_K]^T$ that describes the evolution of the system over time. Thus, the descriptive equation (6.1) is seen as a model of how the observed data have been generated, and a model identification procedure is applied to provide estimates of the coefficients and innovation variances to be used for computing the various frequency domain connectivity measures. In the following, we describe two possible approaches for model identification when the general closed-loop form (6.1) is particularized to the strictly causal MVAR form (6.2) or to the extended MVAR form (6.3).

6.5.1   STRICTLY CAUSAL MVAR IDENTIFICATION

Identification of the strictly causal MVAR model (6.2) can be performed with relative ease by means of estimation methods based on the principle of minimizing the prediction error, that is, the difference between actual and predicted data (see, e.g., the books of Kay (1988) and Lütkepohl (1993) for detailed descriptions). A simple estimator is the vector least-squares method, which is based first on representing Equation 6.2 in compact form as $\mathbf{X} = \mathbf{AZ} + \mathbf{U}$, where $\mathbf{A} = [\mathbf{A}(1) \cdots \mathbf{A}(p)]$ is the $K \times pK$ matrix of the unknown coefficients, $\mathbf{X} = [\mathbf{x}(p+1) \cdots \mathbf{x}(N)]$, and $\mathbf{U} = [\mathbf{u}(p+1) \cdots \mathbf{u}(N)]$ are $K \times (N-p)$ matrices, and $\mathbf{Z} = \left[\mathbf{Z}_1^T \cdots \mathbf{Z}_p^T \right]^T$ is a $pK \times (N-p)$ matrix having $\mathbf{Z}_m = [\mathbf{x}(p-m+1) \cdots \mathbf{x}(N-m)]$ as the mth row block $(m = 1, \ldots, p)$. The method estimates the coefficient matrices through the well-known least-squares formula: $\hat{\mathbf{A}} = \mathbf{XZ}^T \left[\mathbf{ZZ}^T \right]^{-1}$, and the innovation process as the residual time series: $\hat{\mathbf{U}} = \mathbf{X} - \hat{\mathbf{A}}\mathbf{Z}$.

As regards the selection of the model order p, several criteria exist for its determination (see [Lütkepohl, 1993] for a detailed discussion). One common approach is to set the order at the value for which the Akaike figure of merit, defined as $AIC(p) = N \log \det \mathbf{\Sigma} + M^2 p$, reaches a minimum within a predefined range of orders. A helpful hint for practical refinement stands in the consideration that a too low model order would result in the inability to describe essential information about the vector process, while a too high order would bring about overfitting effects implying that noise is captured in the model together with the searched information. While the model identification and order selection methods presented here have good statistical properties, more accurate approaches exist; for example, we refer the reader to Schlögl (2006), for a comparison of different MVAR estimators, and to Karimi (2011), for order selection criteria optimized for MVAR models.

6.5.2   EXTENDED MVAR IDENTIFICATION

While strictly causal MVAR models can be identified by classic regression methods, the identification of extended MVAR models is much less straightforward, because the direction of instantaneous effects, if present, is hard to extract from the data covariance information. In principle, the following approach exploiting two sequential identification stages can be adopted to estimate an extended MVAR model in the form of Equation 6.3:

i. Identify the strictly causal model (6.2) that describes the observed data $x(n), n = 1, \ldots, N$, to estimate the model coefficients, $\hat{A}(l), l = 1, \ldots, p$, and innovations, $\hat{u}(n)$;

ii. Identify the instantaneous model $u(n) = Lw(n)$ that fits the estimated strictly causal innovations $\hat{u}(n)$, to estimate the mixing matrix $\hat{L}$;

iii. Exploit the relations in Equation 6.4 to estimate the extended MVAR coefficients $\hat{B}(0) = I - \hat{L}^{-1}$, $\hat{B}(l) = \hat{L}^{-1}\hat{A}(l), l = 1, \ldots, p$, and the extended innovations, $\hat{w}(n) = \hat{L}^{-1}\hat{u}(n)$.

While the step (i) is easily performed, for example, through the method stated in Section 6.5.1, the step (ii) is complicated by the fact that the instantaneous model may suffer from lack of identifiability, being related to the zero-lag covariance structure of the observed data which is, *per se*, nondirectional. In other words, using only the covariance information, one may find several combinations of L and $w(n)$, which result in the same $u(n)$, and thus describe the observed data $x(n)$ equally well. This problem may be overcome only by exploiting additional information other than that provided by the data covariance alone. The most common way to proceed in this case is to set *a priori* the structure of IC, that is, to impose the direction (though not the strength) of the instantaneous transfer paths. This is achieved by predefining a causal ordering for the observed time series such that no later series causes instantaneously any earlier series, and then solving the instantaneous model through the Cholesky decomposition (see, e.g., Faes and Nollo, 2010b). This solution is efficient and intuitive but has the drawback that it requires the imposition of IC, which can be performed only according to physical considerations, for example, based on the temporal order of the events of interest for each measured variable. While this solution has been applied successfully, for example, in the analysis of cardiovascular variability series (Faes and Nollo, 2010a), it is hard to follow when prior knowledge about the direction of instantaneous effects is not available, as happens commonly in the analysis of neurobiological time series.

A very interesting method for instantaneous model identification has been proposed by Shimizu et al. (2006), who showed how the assumption of non-Gaussianity makes the model identifiable without explicit prior assumption on the existence or nonexistence of given causal effects. This method has been recently combined with classic strictly causal MVAR estimation to perform two-stage extended MVAR model identification as described above by Hyvarinen et al. (2010), who proposed algorithms for efficient estimation of the model and for sparsification and testing of the coefficients, and by Faes et al. (2013), who exploited the model to estimate frequency domain causality and assess patterns of cortical connectivity from multichannel EEG. The method provides a unique solution for the instantaneous model $u(n) = Lw(n)$ under the assumptions that the extended innovations $w(n)$ have *non-Gaussian* marginal distribution, and that the matrix of the instantaneous effects $B(0)$

is *acyclic*, that is, it can be permuted to a lower triangular matrix (which corresponds to assuming that the variables in $\mathbf{w}(n)$ can be arranged in a causal order).

As a first step, the algorithm applies independent component analysis (ICA) on the residuals $\mathbf{u}(n)$. ICA finds a mixing matrix $\mathbf{M}$ and a set of source processes $\mathbf{s}(n)$ such that $\mathbf{u}(n) = \mathbf{M}\mathbf{s}(n)$. The mixing matrix $\mathbf{M}$ can be identified as long as the distribution of $\mathbf{u}(n)$ is a linear, invertible mixture of independent and non-Gaussian components (Shimizu et al., 2006). In fact, the problem is solved looking for the transformation $\mathbf{Q} = \mathbf{M}^{-1}$ that makes the sources $\mathbf{s}(n) = \mathbf{Q}\mathbf{u}(n)$ maximally non-Gaussian, and thus maximally independent. In principle, identification of the instantaneous model could result directly from Independent component analysis (ICA), that is, we could take $\hat{\mathbf{L}} = \hat{\mathbf{M}}$ and $\hat{\mathbf{w}}(n) = \hat{\mathbf{s}}(n)$. Unfortunately, like in most factor-analytic models, order and scaling of the components returned by ICA are completely arbitrary, meaning that the matrices $\mathbf{M}$ and $\mathbf{Q}$ are found up to permutation and scaling. Thus, the second step of the algorithm exploits the desired relationship between $\mathbf{L}$ and $\mathbf{B}(0)$, that is, $\mathbf{L}^{-1} = \mathbf{I} - \mathbf{B}(0)$, to solve the permutation and scaling indeterminacies. Specifically, as $\mathbf{B}(0)$ is expected to have zero diagonal, the estimate of $\mathbf{L}^{-1}$ must have all ones on the main diagonal, and thus the row permutation of $\mathbf{Q} = \mathbf{M}^{-1}$ corresponding to the correct model cannot have a zero on the diagonal. Hence, the matrix $\mathbf{Q}$ resulting from ICA is row permuted to get a matrix $\tilde{\mathbf{Q}}$ with the largest possible elements on the diagonal, and then each row of $\tilde{\mathbf{Q}}$ is divided by its diagonal element to get rescaling toward a new matrix $\bar{\mathbf{Q}}$ with all ones on the diagonal. Finally, the model is solved taking $\hat{\mathbf{L}} = \bar{\mathbf{Q}}^{-1}$. Note that, to get uniqueness of the solution, one has to assume the existence of an instantaneous causal ordering of the observed variables. Indeed, Shimizu et al. (2006) showed that only when $\mathbf{B}(0)$ is acyclic, there is a single ordering of the rows of $\mathbf{Q}$ leading to all nonzero diagonal elements in $\tilde{\mathbf{Q}}$, which can be scaled equal to one in $\bar{\mathbf{Q}}$.

6.5.3 MODEL VALIDATION

Model validation refers to the use of a range of diagnostic tools which are available for checking the adequacy of the estimated model structure. While validation tests are rarely used in practical frequency domain connectivity analysis, they constitute actually fundamental safeguards against drawing erroneous inferences consequently to model misspecification. A striking example is the undervaluation of the impact of instantaneous effects in the assessment of frequency domain causality, which we demonstrated in this chapter and might be avoided by proper checking of the significance of instantaneous correlations in experimental multichannel data. While a thorough description of the statistical tools for model validation goes beyond the scope of this chapter (for detailed descriptions see, e.g., Lütkepohl, 1993), we recall here the model assumptions that must be verified prior to frequency domain analysis.

The basic assumptions which need to be satisfied in MVAR model analysis are those of *whiteness* of the estimated vector innovation process and of *independence* of its scalar components. The assumption of whiteness entails that the innovation variables $w_i(n-l)$ and $w_j(n-m)$ of the extended model (or, equivalently, $u_i(n-l)$ and $u_j(n-m)$ of the strictly causal model) are mutually independent for each $i, j = 1, \ldots, M$ and for each $m \neq l$. The assumption of independence entails that the variables $w_i(n)$ and $w_j(n)$ (or $u_i(n)$ and $u_j(n)$) are mutually independent for each

$i \neq j$. Note that whiteness of the extended innovations $\mathbf{w}(n)$ always corresponds to whiteness of the strictly causal innovations $\mathbf{u}(n)$, because $\mathbf{w}$ and $\mathbf{u}$ differ only in the instantaneous structure. In contrast, independence of $\mathbf{w}$ corresponds to independence of $\mathbf{u}$ only in the absence of IC; in fact, violation of the assumption of independence of the strictly causal innovations is the reason leading to proposing the extended MVAR model in place of the strictly causal one. According to what was explained in Section 6.5.2, when the extended model is implemented, the conditions underlying identifiability of the instantaneous model $\mathbf{u}(n) = \mathbf{L}\mathbf{w}(n)$ have to be verified in addition to the basic assumptions. These conditions consist in the non-Gaussian distribution of the extended innovation process $\mathbf{w}(n)$ and in the acyclicity of the matrix of the instantaneous effects $\mathbf{B}(0)$.

To perform model validation in practical applications, we propose the use of standard statistical tests which are reviewed in Lütkepohl (1993). The whiteness of the model residuals representing estimates of the innovations $\mathbf{u}(n)$ and $\mathbf{w}(n)$ can be tested using the Ljung–Box portmanteau test, which checks for the overall significance of groups of residual correlations. Independence of strictly causal or extended model residuals can be tested by means of Spearman's rho or Kendall's tau tests for independence; rejection of the test for the strictly causal residuals $\hat{\mathbf{u}}(n)$ is taken as an indication of the significance of instantaneous effects among the observed series. If this is the case, the non-gaussianity of the extended residuals $\hat{\mathbf{w}}(n)$ can be tested by means of the Jarque–Bera test for nonnormality of a vector stochastic process, while the acyclicity of the estimated instantaneous effects $\hat{\mathbf{B}}(0)$ can be quantified through the heuristic criterion proposed in Shimizu et al. (2006).

6.5.4 NUMERICAL EXAMPLE

The feasibility of the strictly causal and extended MVAR identification methods was tested on numerical simulations of the illustrative example presented in Section 6.4. To this end, realizations of Equation 6.21 of $N = 500$ samples were obtained generating realizations of the extended innovations $\mathbf{w}(n)$, estimating the corresponding strictly causal innovations as $\mathbf{u}(n) = [\mathbf{I} - \mathbf{B}(0)]^{-1}\mathbf{w}(n)$, and finally feeding a strictly causal model in the form of Equation 6.2 with the estimated $\mathbf{u}(n)$ to get the series $\mathbf{x}(n), n = 1, \ldots, N$. The innovations $w_k(n), k = 1, \ldots, K$, were generated as independent white noises, having either a Gaussian or a non-Gaussian marginal distribution; in the second case, non-gaussianity was achieved first by generating $z_k(n)$ as independent Gaussian white noises and then applying the nonlinear transformation $w_k(n) = \text{sign}(z_k(n))|z_k(n)|^q$, with the exponent q chosen in the range $[0.5, 0.8]$ or $[1.2, 2.0]$ to yield, respectively, a sub-Gaussian or super-Gaussian distribution for w_k.

Figure 6.5 reports the plots of the expected and estimated representative frequency domain connectivity measures under three different scenarios: with non-Gaussian innovations in the absence of IC (Figure 6.5a) and in the presence of IC (Figure 6.5b), and with Gaussian innovations in the presence of IC (Figure 6.5c).

In the absence of IC, all connectivity measures estimated from both the strictly causal and the extended MVAR models fitted to the considered process realization were well adherent to their theoretical profiles (Figure 6.5a); in this case, all LDC measures (gPDC, nPDC, and ePDC) were uniformly zero, indicating the absence

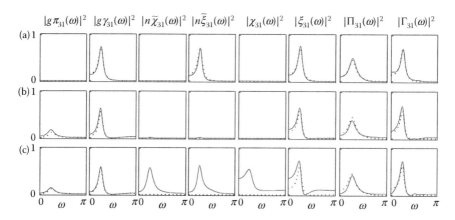

FIGURE 6.5 Theoretical profiles (dotted lines) and estimated profiles (solid lines) of representative connectivity measures from x_1 to x_3 (from left to right: gPDC, gDTF, nPDC, nDTF, ePDC, eDTF, PCoh, Coh) obtained respectively from Equation 6.21 and from a 500 points realization generated under three scenarios: (a) with non-Gaussian innovations and parameter $\delta = 1$; (b) with non-Gaussian innovations and parameter $\delta = 0$; (c) with Gaussian innovations and parameter $\delta = 0$.

of direct causal effects from x_1 to x_3, while the LTC measures (gDTF, nDTF, and eDTF) were nonzero and equal to each other, reflecting the indirect lagged effect from x_1 to x_3 that was imposed by setting $\delta = 1$ in Equation 6.21. Correct estimation was confirmed by the fulfillment of all performed validation tests (i.e., whiteness and independence of $\mathbf{u}(n)$ and $\mathbf{w}(n)$, and non-Gaussianity of $\mathbf{w}(n)$).

The theoretical and estimated profiles of the various connectivity measures were also in good agreement in the presence of IC (see Figure 6.5b). However, in this case, the gPDC derived from the strictly causal MVAR model was nonzero, giving an erroneous indication of direct causality from x_1 to x_3. Rather than wrong model identification, this misleading interpretation is due to the wrong theoretical setting of the strictly causal model (see Section 6.4.2). The unsuitability of the strictly causal representation was evidenced by the fact that the test for independence of the residuals $\mathbf{u}(n)$ was not fulfilled in this case (while all other validation tests were fulfilled).

Finally, we note that when the requirement of non-Gaussian innovations for the extended MVAR model was violated, the connectivity measures derived from the extended representation (nPDC, nDTF, ePDC, eDTF) deviated strongly from their expected values (Figure 6.5c). In particular, the estimated profiles of nPDC and ePDC from x_1 to x_3 are markedly above their corresponding theoretical profiles, thus providing an erroneous indication of lagged and extended causality along a direction over which no causal effects were imposed. This misleading behavior was reflected in model validation by the test for Gaussianity, which revealed the inability of the extended model to yield nonnormally distributed residuals. Note also that in this case with Gaussian innovations the ICA algorithm included in extended MVAR identification may or may not converge for different process realizations. When ICA did not converge, the assumption of independence of $\mathbf{w}(n)$ was also violated; in this case, we also observed that the estimated Coh and PCoh differed markedly from their expected theoretical trend (result not shown in Figure 6.5c).

6.6 CONCLUSIONS

Utilization of the extended MVAR modeling approach for the evaluation of frequency domain connectivity is recommended whenever the interactions among the observed multiple time series cannot be fully explained in terms of lagged effects. In such a case, considering the instantaneous effects changes the coefficients representing the lagged effects as well. We have shown that the presence of significant IC may produce highly misleading patterns of lagged causality, either estimated through the traditional PDC and DTF functions or through their generalized or information extensions, when the classic strictly causal MVAR representation is used. The correct representation of lagged causality, and of extended causality including both lagged and instantaneous effects, is retrieved when frequency domain connectivity is measured from the coefficients of the proposed extended MVAR representation.

Among all measures, ePDC and eDTF should be preferred as they provide a fully meaningful spectral representation respectively of direct and total (direct + indirect) causal effects; indeed, the squared modulus of ePDC and eDTF provides normalized measures of the fraction of inverse spectral power of the input series that is sent to the output series, and of the fraction of spectral power of the output series that is received from the input series, respectively. As an alternative, when IC is void of physical meaning (or when one is interested in lagged effects only), the lagged direct and total causal effects may be quantified in the frequency domain using respectively the lPDC and lDTF, but keeping in mind that they are not normalized, or using respectively the nPDC and nDTF, but keeping in mind that they do not reflect any outgoing inverse power ratio or incoming power ratio. We remark that, when instantaneous effects are negligible, all direct and total causality measures reduce to the gPDC and to the gDTF, respectively, which are thus optimal in this situation. As to the coupling measures, they can be correctly computed indifferently from the strictly causal or the extended MVAR representations to measure direct coupling (PCoh) or total coupling (Coh); however, they cannot elicit directionality of the observed interaction and might indicate a spurious interaction when the two processes under analysis, though not being truly connected, are connected to another common process.

A limitation to the practical utilization of extended MVAR modeling on experimental time series stands in the additional assumptions that need to be set to guarantee identifiability of the model. While the reported identification algorithm has been proven to efficiently estimate instantaneous and lagged model coefficients when the assumptions of non-Gaussianity and acyclicity are met, one should be aware that failing to fulfill these assumptions may lead to wrong coefficient estimates and thus to misleading patterns of frequency domain connectivity. When the innovations are Gaussian, independent component analysis is likely unable to estimate the instantaneous model, and this may result in estimating arbitrary connectivity patterns (Faes et al., 2013); when an instantaneous causal ordering does not exist for the observed time series, the matrix of the instantaneous effects is not acyclic and the solution is not unique (Hyvarinen et al., 2010). While alternative identification approaches to relax these assumptions are under investigation (Hoyer et al., 2008; Lacerda et al., 2008), we remark on the importance of model validation, which can be purposefully exploited to test the necessity of resorting to extended MVAR modeling and of checking the underlying requirements directly from the estimated model residuals.

ACRONYMS

Coh	Coherence
Coup	Coupling
DCoup	Direct coupling
DTF	Directed transfer function
eDTF	extended directed transfer function
ePDC	extended partial directed coherence
gDTF	generalized directed transfer function
gPDC	generalized partial directed coherence
iDTF	information directed transfer function
iPDC	information partial directed coherence
lDTF	lagged directed transfer function
lPDC	lagged partial directed coherence
nDTF	normalized lagged directed transfer function
nPDC	normalized lagged partial directed coherence
EDC	Extended direct causality
ETC	Extended total causality
FT	Fourier's transform
IC	Instantaneous causality
ICA	Independent component analysis
LDC	Lagged direct causality
LTC	Lagged total causality
MVAR	Multivariate autoregressive
PCoh	Partial coherence
PDC	Partial directed coherence

REFERENCES

Astolfi, L., F. Cincotti, D. Mattia et al. 2006. Assessing cortical functional connectivity by partial directed coherence: Simulations and application to real data. *IEEE Trans. Biomed. Eng.* 53: 1802–1812.

Baccalá, L. A., D. Y. Takahashi, and K. Sameshima. 2007. Generalized partial directed coherence. In *15th International Conference on Digital Signal Processing* (DSP2007), 163–166, Cardiff, Wales, UK.

Baccalá, L. A. and K. Sameshima. 2001. Partial directed coherence: A new concept in neural structure determination. *Biol. Cybern.* 84: 463–474.

Baccalá, L. A., K. Sameshima, G. Ballester et al. 1998. Studying the interaction between brain structures via directed coherence and Granger causality. *Appl. Signal Process. 5*: 40–48.

Brillinger, D. R., J. H. L. Bryant, and J. P. Segundo. 1976. Identification of synaptic interactions. *Biol. Cybern. 22*: 213–228.

Dahlhaus, R. 2000. Graphical interaction models for multivariate time series. *Metrika 51*: 157–172.

Eichler, M. 2006. On the evaluation of information flow in multivariate systems by the directed transfer function. *Biol. Cybern. 94*: 469–482.

Eichler, M., R. Dahlhaus, and J. Sandkuhler. 2003. Partial correlation analysis for the identification of synaptic connections. *Biol. Cybern. 89*: 289–302.

Faes, L., S. Erla, and G. Nollo. 2012. Measuring connectivity in linear multivariate processes: Definitions, interpretation, and practical analysis. *Comput. Math. Method. Med. 2012*: 140513.

Faes, L., S. Erla, A. Porta et al. 2013. A framework for assessing frequency domain causality in physiological time series with instantaneous effects. *Philos. Transact. A Math. Phys. Eng. Sci.*, 371:20110618 (21 pages).

Faes, L., and G. Nollo. 2010a. Assessing frequency domain causality in cardiovascular time series with instantaneous interactions. *Methods Inform. Med. 49*: 453–457.

Faes, L. and G. Nollo. 2010b. Extended causal modelling to assess partial directed coherence in multiple time series with significant instantaneous interactions. *Biol. Cybern. 103*: 387–400.

Gevers, M. R. and B. D. O. Anderson. 1981. Representations of jointly stationary stochastic feedback processes. *Int. J. Control 33*: 777–809.

Granger, C. W. J. 1969. Investigating causal relations by econometric models and cross-spectral methods. *Econometrica 37*: 424–438.

Granger, C. W. J. 1980. Testing for causality: A personal viewpoint. *J. Econom. Dynam. Control 2*: 329–352.

Hoyer, P. O., A. Hyvarinen, R. Scheines et al. 2008 Causal discovery of linear acyclic models with arbitrary distributions. *Proceedings of the Twenty-Fourth Conference on Uncertainty in Artificial Intelligence*, 282–289. Helsinki, Finland.

Hyvarinen, A., K. Zhang, S. Shimizu et al. 2010. Estimation of a structural vector autoregression model using non-gaussianity. *J. Mach. Learn. Res. 11*: 1709–1731.

Kaminski, M., M. Ding, W. A. Truccolo et al. 2001. Evaluating causal relations in neural systems: Granger causality, directed transfer function and statistical assessment of significance. *Biol. Cybern. 85*: 145–157.

Karimi, M. 2011. Order selection criteria for vector autoregressive models. *Signal Process. 91*: 955–969.

Kay, S. M. 1988. *Modern Spectral Estimation. Theory and Application*. Englewood Cliffs: Prentice-Hall.

Lacerda, G., P. Spirtes, J. Ramsey et al. 2008. Discovering cyclic causal models by independent component analysis. *Proceedings of the Twenty-Fourth Conference on Uncertainty in Artificial Intelligence*, 366–374. Helsinki, Finland.

Lütkepohl, H. 1993. *Introduction to Multiple Time Series Analysis*. Berlin: Springer.

Schlogl, A. 2006. A comparison of multivariate autoregressive estimators. *Signal Process. 86*: 2426–2429.

Shimizu, S., P. O. Hoyer, A. Hyvarinen et al. 2006. A linear non-gaussian acyclic model for causal discovery. *J. Mach. Learn. Res. 7*: 2003–2030.

Takahashi, D. Y., L. A. Baccalá, and K. Sameshima. 2010. Information theoretic interpretation of frequency domain connectivity measures. *Biol. Cybern. 103*: 463–469.

Whittaker, J. 1990. *Graphical Models in Applied Multivariate Statistics*. Chichester: Wiley.

Winterhalder M., B. Schelter, W. Hesse et al. 2005. Comparison directed of linear signal processing techniques to infer interactions in multivariate neural systems. *Signal Process. 85*: 2137–2160.

7 Asymptotic PDC Properties

Koichi Sameshima, Daniel Y. Takahashi, and Luiz A. Baccalá

CONTENTS

7.1 INTRODUCTION

This chapter aims to provide an overview of the statistical properties of *partial directed coherence* inferential use; see Chapters 4 and 5 for PDC for definitions. Whereas objective time-domain criteria for detecting Granger causality-based connectivity abound (Geweke, 1982, 1984; Geweke et al., 1983; Lütkepohl, 1993; Baccalá et al., 1998) (see also Chapter 3), effective frequency domain tests are more recent and have only slowly followed the introduction of frequency domain methods. Early characterization frequency domain statistical behavior made extensive use of Monte Carlo simulations revealing that their statistical variability is a function of the signal's frequency content (Baccalá and Sameshima, 2001).

Early attempts to provide objective means of frequency domain inference were pioneered by Kamiński et al. (2001) and employed spectral phase resampling methods (Prichard and Theiler, 1994), which were later further elaborated in Baccalá et al. (2006) in the connectivity context by comparing phase resampling to more traditional prediction error-based methods as described by Lütkepohl (1993) showing their essential equivalence in many cases.

Despite the usefulness of the later computer-intensive techniques, deeper understanding of the statistical factors remained amiss and only started to change with the work of Schelter et al. (2005), who published the first PDC asymptotic behavior results.

Roughly speaking, regardless of whether it is carried out in the time domain or in the frequency domain, the statistical connectivity inference problem involves two very distinct issues:

1. Detecting the presence of significant connectivity at a given frequency—the connectivity *detection* problem;
2. Determining the confidence interval of the estimated value when it is significant at that frequency—the connectivity *quantification* problem.

While Schelter et al. (2005) examined only issue 1, Takahashi et al. (2007) showed that both issues could be rigorously examined from a single asymptotic statistical point of view.

The availability of single trial-based confidence intervals allows the consistent comparison of connectivity strengths under different experimental conditions without having to resort to ANOVA-based repeated experiments for inference (Wang et al., 2008).

At this point, it is perhaps interesting to note that early attempts at applying PDC were mainly hampered in practice by the nonexistence of a clear understanding of how its originally defined form depended on factors such as signal length, its spectrum, and the number of simultaneously analyzed time series. The latter issues were in part responsible, in the mean time, for the need to consider other forms of PDC in Baccalá et al. (2007) and Takahashi et al. (2010) (see Chapter 4) naturally compounding the inference problem even more. Only recently has it been possible to deduce the asymptotics of the latter PDC forms from a single unified perspective (Baccalá et al., 2013) whose results are briefly reviewed and further illustrated here.

To simplify the exposition, most mathematical details are presented without proof. For completeness, the main line of mathematical development is presented in the Appendix which the accompanying code mirrors very closely, having been written for clarity. A more efficient code implementation is in progress.

The notation used here comes directly from Chapter 4 and is consistent with Baccalá et al. (2013).

7.2   ASYMPTOTIC STATISTICAL INFERENCE

7.2.1   BASIC RATIONALE

A prerequisite for the validity of the results presented here is that the data have been successfully fit by a vector autoregressive model (4.1) of appropriate order (see Chapters 3 and 4 for some related model diagnosis issues).

In that case, under mild assumptions (Lütkepohl, 2005), as the number of observed data points n_s increases, both the model parameters $a_{ij}(k)$ and estimated prediction error variances σ_{ij} tend to become normally distributed. This, together with the realization that PDC is a continuous function of the latter quantities, allows expanding its statistical distribution as a Taylor series based on the model parameter statistics whose higher-order terms become quantatively less important as n_s increases. This expansion procedure is known as the *Delta Method* (van der Vaart, 1998).

The main unified result described in Baccalá et al. (2013) is that significant PDC values are asymptotically Gaussian, a property that is inherited from the asymptotics of the parameters regardless of the PDC brand.

In Baccalá et al. (2013), we also show that PDC gaussianity breaks down when connectivity is absent, in which case the next term in the Delta Method provides the applicable asymptotic distribution that can then be used to construct a rigorous hypothesis test for connectivity.

7.2.2 Results in Brief

More formally, the true absolute squared value of PDC, from j to i, at the frequency λ may be written as the following ratio:

$$|\pi_{ij}(\lambda)|^2 = \pi(\boldsymbol{\theta},\lambda) = \frac{\pi_n(\boldsymbol{\theta},\lambda)}{\pi_d(\boldsymbol{\theta},\lambda)}, \qquad (7.1)$$

where $\boldsymbol{\theta}$ vector collects a_{ij} and σ_{ij} dependence according to the desired PDC form.

Then the following results hold:

1. *Confidence Intervals*   For large n_s, estimated (7.1) is asymptotically normal, that is,

$$\sqrt{n_s}\left(|\widehat{\pi}_{ij}(\lambda)|^2 - |\pi_{ij}(\lambda)|^2\right) \to \mathcal{N}(0,\gamma^2(\lambda)), \qquad (7.2)$$

   where $\gamma^2(\lambda)$ is a frequency-dependent variance whose form depends on the PDC brand of interest.

2. *Null Hypothesis Threshold*   Under the null hypothesis,

$$H_0 : \left|\pi_{ij}(\lambda)\right|^2 = 0 \qquad (7.3)$$

$\gamma^2(\lambda)$ vanishes identically and Equation 7.2 no longer holds. The asymptotic behavior of Equation 7.1 is dominated by the next Delta Method expansion term with an $O(n_s^{-1})$ dependence. This term is quadratic in $\boldsymbol{\theta}$ and leads to a distribution that can be written in terms of the distribution of a linear combination of two properly weighted χ_1^2. The weights of the linear combination are frequency dependent.

This can be formally summarized as

$$n_s\,\pi_d(\boldsymbol{\theta},\lambda)\left(|\widehat{\pi}_{ij}(\lambda)|^2 - |\pi_{ij}(\lambda)|^2\right) \overset{d}{\to} \sum_{k=1}^{q} l_k(\lambda)\chi_1^2. \qquad (7.4)$$

It is possible to show that the $l_k(\lambda)$ weights depend only on PDC's numerator $\pi_n(\boldsymbol{\theta},\lambda)$ and that PDC denominator dependence is implicit on the left-hand side of Equation 7.4.

Specific numerical integration methods are needed to obtain quantile values from the distribution that results from the linear combination of χ_1^2 variables (Patnaik,

1949; Imhof, 1961). From them, one may compute adequate hypothesis test thresholds at the desired significance level (Takahashi et al., 2007). Implementation of the latter is provided in the accompanying software.

7.3 NUMERICAL ILLUSTRATIONS

We now briefly move on to consider some examples and start by applying the asymptotic results to controlled simulation contexts to provide a feel for performance (Section 7.3.1). This is followed by examples involving real data (Section 7.3.2). Additional illustrations may be found in Takahashi et al. (2007), de Brito et al. (2010), and Baccalá et al. (2013).

7.3.1 SIMULATED MODELS

We begin the simplest possible simulation example:

EXAMPLE 7.1: SIMPLE MODEL

Let

$$
\begin{cases}
x_1(n) = 0.5x_1(n-1) + w_1(n) \\
x_2(n) = 0.5x_2(n-1) + 0.5x_1(n-1) + w_2(n)
\end{cases}
\tag{7.5}
$$

with independent unit variance zero mean Gaussian $w_i(n)$ innovations noise. Figure 7.1 shows original PDC results for as few as $n_s = 100$ at a significance level $\alpha = 1\%$, obtained by running the **simple_pdc_asymp.m** routine from the AsympPDC package in the accompanying CD.

The reader may compare it to the allied bootstrap-based inference in Baccalá et al. (2006); a fuller systematic comparison is being carried out elsewhere.

EXAMPLE 7.2: LOOP OSCILLATOR

Next consider:

$$
\begin{cases}
x_1(n) = 0.95\sqrt{2}x_1(n-1) - 0.9025x_1(n-2) \\
\qquad\quad + 0.35x_3(n-1) + w_1(n) \\
x_2(n) = 0.5x_1(n-1) + 0.5x_2(n-1) + w_2(n) \\
x_3(n) = -0.5x_3(n-1) + x_2(n-1) + w_3(n),
\end{cases}
\tag{7.6}
$$

where connectivity is characterized by positive feedback due to the $0.35x_3(n-1)$ term back into the oscillator represented by $x_1(n)$.
Choosing

$$
\Sigma_w = \begin{bmatrix} 1 & 5 & 0.3 \\ 5 & 100 & 2 \\ 0.3 & 2 & 1 \end{bmatrix}
\tag{7.7}
$$

makes ιPDC a natural choice for analysis whose behavior for an $n_s = 2000$ long realization is shown in Figure 7.2 in standard form showing correct connectivity

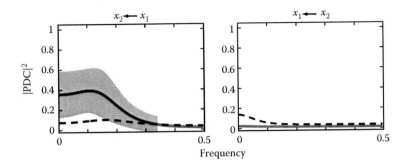

FIGURE 7.1 Original PDC connectivity results for Example 7.1. Gray shade represents the 99% confidence intervals associated with significant values at $\alpha = 1\%$ whose decision thresholds are indicated by dashed lines. Gray lines indicate nonsignificant PDC values.

detection at $\alpha = 5\%$. Note how narrow are the shades portraying the confidence intervals for values under null hypothesis rejection. The graphs portray a single output realization from **loop_ipdc_asymp.m**.

Monte Carlo behavior at $\lambda = 0.25$ for 2000 realizations is displayed in Figure 7.3 as a function of $n_s = \{100, 500, 1000, 2000\}$ confirming the statistical goodness-of-fit improvement for both existing, $x_2 \rightarrow x_3$ (normally distributed, see Equation 7.2), and nonexisting connections, $x_3 \rightarrow x_2$ (following Equation 7.4). The readers may execute **loop_monte_carlo_main.m** to confirm these results. □

EXAMPLE 7.3: MIDBAND UNBALANCED NOTCH

Data for this example were generated using the following equations:

$$\begin{cases} x_1(n) = 0.5\, x_1(n-1) + w_1(n) \\ x_2(n) = x_1(n-1) + x_1(n-3) + 0.5x_2(n-1) + w_2(n), \end{cases} \tag{7.8}$$

which are meant to impose *null* information transmission at midband $\lambda = 0.25$ through the notch filter term $x_1(n-1) + x(n-3)$. PDC and ιPDC (Figure 7.4) have rather different overall qualitative behavior given the power balance built into the simulated innovations covariance:

$$\Sigma_w = \begin{bmatrix} 100 & 0.5 \\ 0.5 & 1 \end{bmatrix}. \tag{7.9}$$

Yet close examination of details in Figure 7.4a,b shows successful detection of connectivity lack at $\lambda = 0.25$ whose behavior is confirmed by the Monte Carlo results in Figure 7.5. □

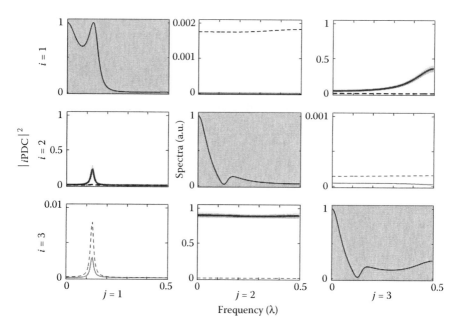

FIGURE 7.2 Standard form of ιPDC display using $n_s = 2000$ and $p = 2$ where solid lines are surrounded by gray-shaded 95% confidence intervals; dashed lines indicate ιPDC critical thresholds at $\alpha = 5\%$. As in Figure 7.1, the gray lines indicate the below threshold ιPDC values. (The graph is an edited version generated by the **loop_ipdc_asymp.m** in the package.)

EXAMPLE 7.4: RESONANT LINK

As a last simulated illustration, consider the case where a connection is made via a bandpass filter embodied in the $y(n)$ variable, which is not measured directly:

$$\begin{cases} x_1(n) = -0.5x_1(n-1) + w_1(n) \\ y(n) = 0.95\sqrt{2}y(n-1) - 0.9025y(n-2) + x_1(n-1) \\ x_2(n) = -0.5x_2(n-1) + y(n-1) + w_2(n), \end{cases} \quad (7.10)$$

where only $x_1(n)$ and $x_2(n)$ are subject to analysis and where their coupling is made through $y(n)$ with

$$\Sigma_w = \begin{bmatrix} 2 & 5 \\ 5 & 20 \end{bmatrix}. \quad (7.11)$$

VAR model estimates based on the restricted set $[x_1(n)\,x_2(n)]^T$ generated via Equation 7.10 usually lead to very high model orders since the actual parsimonious model is VARMA (vector autoregressive moving average) when $y(n)$ is not measured and is equivalent to VAR(∞) models. Despite this, reasonable connectivity estimates are attained (Figure 7.6). It is interesting to note that performance improvement can be obtained if n_s is replaced by $n_s - \hat{p}$ where $\hat{p}$ is the model order estimated based on the data (Takahashi et al., 2008). $\quad\square$

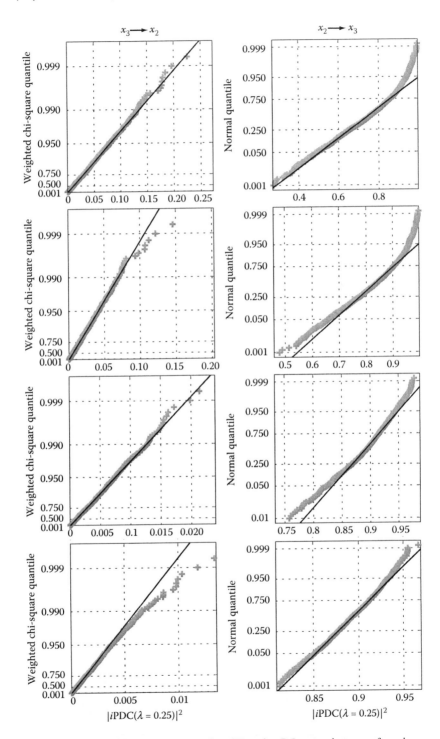

FIGURE 7.3 Monte Carlo estimation results of Equation 7.6 system in terms of varying $n_s = \{100, 500, 1000, 2000\}$ showing the statistical convergence of ιPDC's estimator at $\lambda = 0.25$ for nonexisting, $x_3 \rightarrow x_2$ (left panels) and existing, $x_2 \rightarrow x_3$ (right-panels) connections. These graphs were generated by **loop_monte_carlo_main.m** and edited later. The model order was

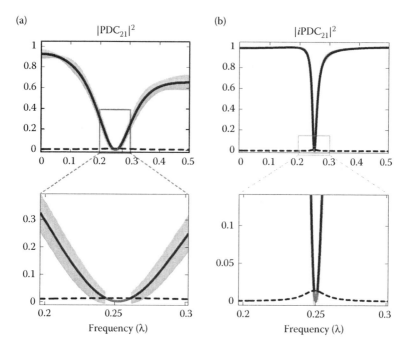

FIGURE 7.4 Comparison between PDC (a) and ιPDC (b) behavior around the notch frequency ($\lambda = 0.25$). The graphs were generated by **notch_pdc_ipdc_asymp.m** and later edited. The model order was set to the correct value $p = 3$ for $\alpha = 5\%$.

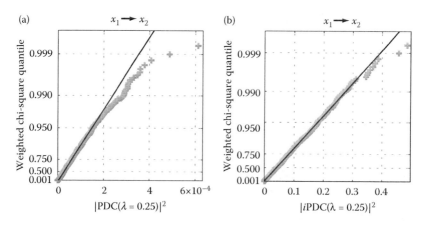

FIGURE 7.5 Example 7.2 Monte Carlo results of the nonsignificant connection $x_1 \rightarrow x_2$ for PDC (a) and ιPDC (b) at $\lambda = 0.25$. The graphs were generated by **notch_monte_carlo_main.m** and edited later. The model order was set to the correct value $p = 2$.

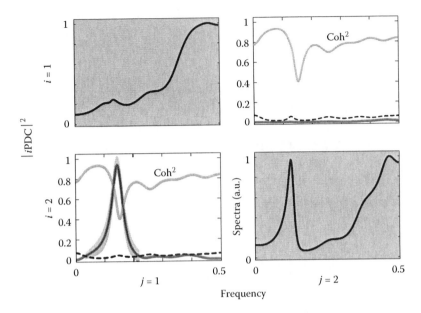

FIGURE 7.6 Standard ιPDC display for Example 7.4 (generated with **resonant_ ipdc_asymp.m**) showing the spectra for both time series $x_1(n)$ and $x_2(n)$ along the main diagonal. The unidirectional resonant influence $x_1(n) \to x_2(n)$ is clearly seen ($\alpha = 0.05$) . The graph also depicts the ordinary coherence between the series (light gray curve). Ordinary coherence's low value around the transmission frequency stems from the large built-in innovations covariance adopted for the example.

7.3.2 REAL DATA

In the following, we consider a set of examples geared for the three PDC modalities discussed in Chapters 4 and 5, respectively.

EXAMPLE 7.5: SUNSPOT–MELANOMA TIME SERIES

In this example, we show that for the mere $n_s = 37$ observations of the Sunspot-Melanoma data (see Example 4.2), when the thresholds and confidence intervals are taken into account Figure 7.7, even the original PDC furnishes the directional inference as to the detection problem—compare with Figure 4.1a. In Figure 7.7 the computed PDC is above the threshold for the Sunspot $\rightarrow$ Melanoma time series, and hence its connectivity is significant, despite the figure's awkwardness. In the opposite direction, the decision threshold is above 1, and hence the data furnish no evidence of Melanoma $\rightarrow$ Sunspot causality as expected.

The situation is clearly improved by using gPDC where both the intuition of the graphical representation and test results agree (see Figure 7.8). □

Now for something different.

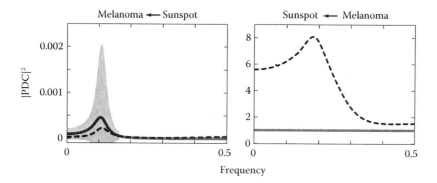

FIGURE 7.7 Standard form display of Sunspot–Melanoma PDC showing that despite the estimator response size problems due to innovations covariance imbalances, correct inference results at $\alpha = 1\%$ (**sunspot_melanoma_pdc_asymp.m** was used to generate this and Figure 7.8). Compare with Figure 4.1a.

EXAMPLE 7.6: EL NIÑO VERSUS SOUTHERN OSCILLATION INDEX BEHAVIOR

Consider the climatological time series involving El Niño data and the Southern Oscillation Index (SOI) spanning 45 years from 1962 (Figure 7.9) whose ιPDC results are shown in Figure 7.10 where the El Niño monthly temperatures are affected by SOI activity at low frequencies whereas the reciprocal influence is marginal. This unidirectional activity from SOI to El Niño is further confirmed by an ordinary time-domain Granger causality test ($p < 10^{-4}$) (**meteodata_pdc_asymp.m** in the package).

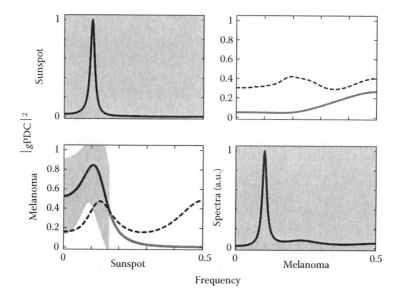

FIGURE 7.8 Standard form display of Sunspot–Melanoma gPDC results ($\alpha = 1\%$) (also created by **sunspot_melanoma_pdc_asymp.m.**).

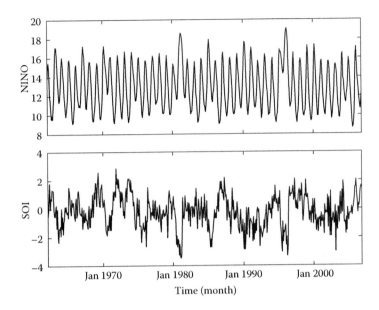

FIGURE 7.9 The monthly data for Example 7.6 comprise 45 years from 1962 NINO34 observations (El Niño average temperature above and below 25°C between ±5 degrees of latitude and between 170 and 120 W degrees of longitude) and SOI – the average pressure difference between Darwin and Tahiti both as available from http://www.esrl.noaa.gov/psd/gcos_wgsp/Timeseries/. This data set is provided in the CD as the **meteodata.txt** ASCII-file with channel identification file, **meteodata.lbl**.

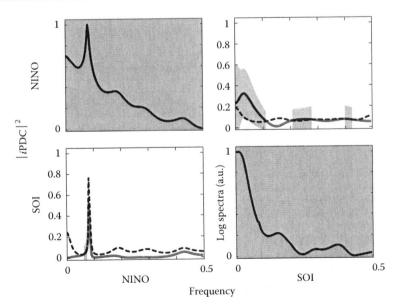

FIGURE 7.10 *i*PDC analysis of Figure 7.9 data showing low-frequency influence from SOI onto the NINO Series. The AIC model order criterion was chosen leading to $p = 9$. The very small effect of NINO toward SOI is borderline and probably due to modeling errors ($\alpha = 1\%$). This analysis was carried out by running **meteodata_ipdc_asymp.m**.

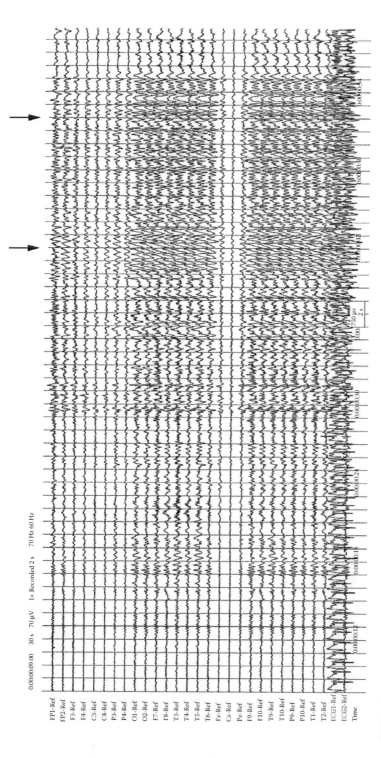

FIGURE 7.11 The final ictal portion of 10-20 EEG signal traces was subject to *g*PDC analysis as shown between the arrow marks for Example 7.7. Model estimates were computed over a 10 s epoch under a sampling frequency of 200 Hz. The data are available in **ictus_data.mat** in the package.

$|g\text{PDC}|^2$

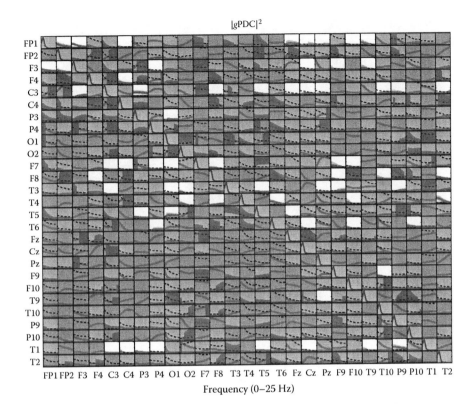

Frequency (0–25 Hz)

FIGURE 7.12   (See color insert.) gPDC results for the ictal EEG episode of Example 7.7. Significant connectivities at $\alpha = 1\%$ are shown in red. White squares point to $|g\text{PDC}|^2 > 0.1$, whereas blue squares refer to values between 0.1 and 0.01 and purple squares are below the latter level. This graph is an edited output of the **ictus_gpdc_analysis.m** routine provided in the CD.

Note that ιPDC was used to face the unbalanced nature of the estimated prediction error covariance matrix:

$$\hat{\boldsymbol{\Sigma}}_W = \begin{bmatrix} .05 & -.01 \\ -.01 & .5 \end{bmatrix}. \tag{7.12}$$

The ENSO (El Niño Southern Oscillation) phenomenon has been the object of a great many recent studies and involves nonlinear fluid dynamic models (Clarke, 2008; Sarachik and Cane, 2010) and its discussion is beyond our intended scope. The example's intent is to show that usual analyses based on correlation may be expanded to encompass frequency domain causality representations.

Using **meteodata_pdc_asymp.m** in the package may be helpful in portraying some of the practical difficulties the reader may find in processing actual data, especially when the phenomena involved may contain substantial nonlinearities and episodic nonstationarities. Note that the VAR models obtained do not strictly pass all model quality checks.   □

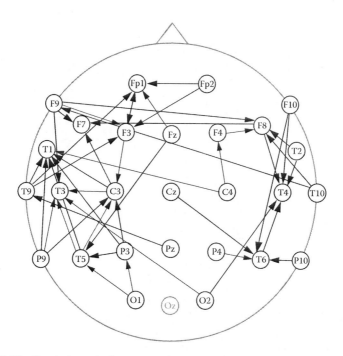

FIGURE 7.13 Graph theoretical representation of the ictal period analysis is described in Example 7.7 for significant connections ($\alpha = 1\%$) with gPDC squared above 0.1.

EXAMPLE 7.7: ANALYSIS OF 27 EEG SIMULTANEOUS CHANNELS OVER AN ICTAL EPISODE

The computed gPDC standard form results comprising 27 channels of the ictal episode portrayed in Figure 7.11 are shown in Figure 7.12 in the 0–25 Hz range. Significant values are marked in red while background colors call attention to gPDC value ranges (purple up to 0.01, blue up to 0.1, and white above the latter). The directed graph (Figure 7.13) sums up the active (above threshold dashed line) connections for $|gPDC|^2 > 0.1$.

The data refer to a clinically confirmed left temporal mesial data case and form part of the work currently in progress. The data segment shown in Figure 7.11 is provided in the **ictus_data.mat** file while Figure 7.12 is an edited version of the output obtained by running **ictus_gpdc_analysis.m**. For the reader's convenience, the result variables are provided in the **ictus_gpdc_analysis_result.mat** with further details in **ictus_gpdc_analysis.log**, which contains the output listing generated by the PDC analysis routine.   □

In the above examples, the evaluation of model quality is the really important limiting factor to successful inference whether one wishes to compare absolute PDC values or the null hypothesis of interaction existence.

7.4   CONCLUSIONS AND FUTURE WORK

This chapter's intent has been to introduce the reader to the use of asymptotic results regarding various PDC forms and show their essential equivalence when null-hypothesis tests for connectivity are concerned. The package provided with the chapter gives the reader the chance to experiment with toy models and his own data. Current work is under way to compare the present rigorous asymptotic statistics with the currently popular resampling-based approaches.

Examples outside the realm of neuroscience were included that show the wider applicability of the results.

APPENDIX: MATHEMATICAL DETAILS

This material is included here for completeness and may be skipped without loss. It is a brief sketch of how Equations 7.2 and 7.4 are obtained. Deeper technical details can be obtained from Takahashi et al. (2007) and Baccalá et al. (2013). Furthermore, the software package included herein is a direct reader-friendly implementation of the current developments. Notation closely follows that of Chapter 4 and is compatible with the one adopted in Baccalá et al. (2013).

In connection to a properly fitted VAR model (4.1), the cornerstone result is associated with the behavior of $a_{ij}(r)$ and σ_{ij} which may be reduced to vectors by column stacking their respective matrices:

$$\boldsymbol{\alpha} = vec[\mathbf{A}(1)\,\mathbf{A}(2)\cdots\mathbf{A}(p)] \tag{7.13}$$

and, similarly, the residual noise covariance matrix $\boldsymbol{\sigma} = vec\,\boldsymbol{\Sigma}_\mathbf{w}$ to produce the parameter vector $\boldsymbol{\theta}^T = \begin{bmatrix} \boldsymbol{\alpha}^T \boldsymbol{\sigma}^T \end{bmatrix}^T$, whose asymptotic behavior is normal (Lütkepohl, 2005):

$$\sqrt{n_s}(\hat{\boldsymbol{\theta}} - \boldsymbol{\theta}) \to \mathcal{N}(0, \boldsymbol{\Omega}_\theta), \tag{7.14}$$

where its covariance

$$\boldsymbol{\Omega}_\theta = \begin{bmatrix} \boldsymbol{\Omega}_\alpha & \mathbf{0} \\ \mathbf{0} & \boldsymbol{\Omega}_\sigma \end{bmatrix} \tag{7.15}$$

is readily computable using the vector time series $\mathbf{x}(n)$ as $\boldsymbol{\Omega}_\alpha = \boldsymbol{\Gamma}_\mathbf{x}^{-1} \otimes \boldsymbol{\Sigma}_\mathbf{w}$, where $\boldsymbol{\Gamma}_\mathbf{x} = E[\bar{\boldsymbol{x}}(n)\bar{\boldsymbol{x}}^T(n)]$ for

$$\bar{x}(n) = [x_1(n)\cdots x_K(n)\cdots$$
$$x_1(n-p+1)\cdots x_K(n-p+1)]^T; \tag{7.16}$$

and

$$\boldsymbol{\Omega}_\sigma = 2\mathbf{D_K}\mathbf{D_K^+}(\boldsymbol{\Sigma}_\mathbf{w} \otimes \boldsymbol{\Sigma}_\mathbf{w})\mathbf{D}_K^{+T}\mathbf{D}_K^T, \tag{7.17}$$

where $\mathbf{D}_K^+$ is the Moore–Penrose pseudo-inverse of the standard duplication matrix. An estimated version of $\boldsymbol{\Sigma}_\mathbf{w}$ obtained after model fitting furnishes Equation 7.17.

The conditions for the validity of Equation 7.14 are associated with the finitude of the higher-order statistical cross moments of the $\mathbf{x}(n)$ time-series data (Lütkepohl,

2005). Also note the asymptotic independence between $a_{ij}(r)$ and σ_{ij} embodied by the block-diagonal character of (7.15).

All PDC definitions involve the frequency λ whose effect may be included considering the change of the following variables from $\boldsymbol{\alpha}$ to $\mathbf{a}(\lambda)$

$$\mathbf{a}(\lambda) = \begin{bmatrix} \mathrm{vec}(\mathrm{Re}(\bar{\mathbf{A}}(\lambda))) \\ \mathrm{vec}(\mathrm{Im}(\bar{\mathbf{A}}(\lambda))) \end{bmatrix} = \begin{bmatrix} \mathrm{vec}(\mathcal{I}_{pK^2}) \\ \mathbf{0} \end{bmatrix} - \mathcal{C}(\lambda)\boldsymbol{\alpha}, \tag{7.18}$$

where $\mathcal{I}$ is the $N \times N$ identity matrix and

$$\mathcal{C}(\lambda) = \begin{bmatrix} \mathbf{C}(\lambda) \\ -\mathbf{S}(\lambda) \end{bmatrix},$$

whose blocks are the $K^2 \times pK^2$ dimensional of the form

$$\mathbf{C}(\lambda) = [\mathbf{C}_1(\lambda) \cdots \mathbf{C}_p(\lambda)] \quad \mathbf{S}(\lambda) = [\mathbf{S}_1(\lambda) \cdots \mathbf{S}_p(\lambda)],$$

for

$$\mathbf{C}_r(\lambda) = \mathrm{diag}([\cos(2\pi r\lambda) \cdots \cos(2\pi r\lambda)]), \quad \mathbf{S}_r(\lambda) = \mathrm{diag}([\sin(2\pi r\lambda) \cdots \sin(2\pi r\lambda)]).$$

Under these conditions, $\mathbf{a}(\lambda)$ is also asymptotically normal and its covariance is trivially given by

$$\boldsymbol{\Omega}_{\mathbf{a}}(\lambda) = \mathcal{C}(\lambda)\boldsymbol{\Omega}\mathcal{C}^T(\lambda) \tag{7.19}$$

for

$$\boldsymbol{\Omega} = \begin{bmatrix} \boldsymbol{\Omega}_\alpha & \boldsymbol{\Omega}_\alpha \\ \boldsymbol{\Omega}_\alpha & \boldsymbol{\Omega}_\alpha \end{bmatrix},$$

From $\mathbf{a}(\lambda)$, one can readily express all PDC forms using:

$$\pi_n(\boldsymbol{\theta}, \lambda) = \mathbf{a}(\lambda)^T \mathbf{I}_{ij}^c \boldsymbol{S}_n(\boldsymbol{\sigma}) \mathbf{I}_{ij}^c \mathbf{a}(\lambda) \tag{7.20}$$

and

$$\pi_d(\boldsymbol{\theta}, \lambda) = \mathbf{a}(\lambda)^T \mathbf{I}_j^c \boldsymbol{S}_d(\boldsymbol{\sigma}) \mathbf{I}_j^c \mathbf{a}(\lambda) \tag{7.21}$$

that is, where both the numerator and denominator in Equation 7.1 are quadratic forms in $\mathbf{a}(\lambda)$ and where $\boldsymbol{S}_n(\boldsymbol{\sigma})$ and $\boldsymbol{S}_d(\boldsymbol{\sigma})$ take on specific values according to PDC type (Table 7.1).

The $\mathbf{I}_{ij}^c$ and $\mathbf{I}_j^c$ above are responsible for selecting the i, j pairs of interest and are defined as

$$\mathbf{I}_{ij}^c = \begin{bmatrix} \mathbf{I}_{ij} & 0 \\ 0 & \mathbf{I}_{ij} \end{bmatrix}, \quad \mathbf{I}_j^c = \begin{bmatrix} \mathbf{I}_j & 0 \\ 0 & \mathbf{I}_j \end{bmatrix}.$$

where

1. $\mathbf{I}_{ij}$ whose entries are zero except for indices of the form $(l, m) = ((j - 1)K + i, (j - 1)K + i)$, which equal 1.

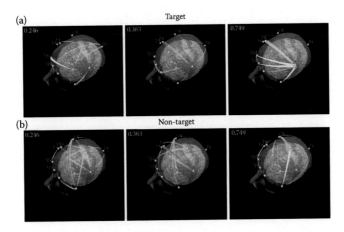

FIGURE 2.7 Snapshots from the movie presenting significant changes in transmissions in one subject, for target (a) and nontarget (b). Intensity of flow changes for increase: from pale yellow to red, for decrease: from light to dark blue. In the right upper corner, the time after cue presentation (in seconds). (Reprinted from Blinowska et al. 2010. *Brain Topogr. 23*: 205–213.)

$|gPDC|^2$

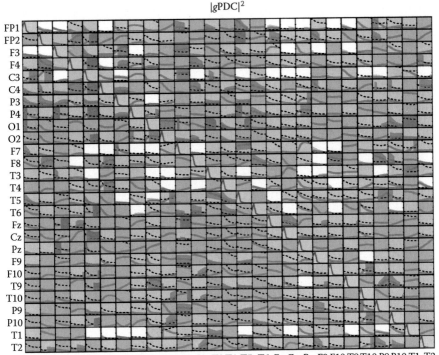

Frequency (0–25 Hz)

FIGURE 7.12 *g*PDC results for the ictal EEG episode of Example 7.7. Significant connectivities at $\alpha = 1\%$ are shown in red. White squares point to $|gPDC|^2 > 0.1$, whereas blue squares refer to values between 0.1 and 0.01 and purple squares are below the latter level. This graph is an edited output of the **ictus_gpdc_analysis.m** routine provided in the CD.

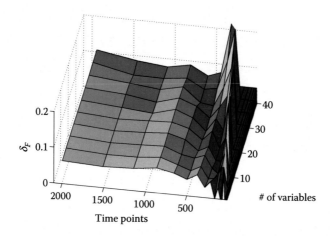

FIGURE 8.5 Granger-causal relationships of a given pair of variables in a nonlinear AR model for a different number of variables and time-series length.

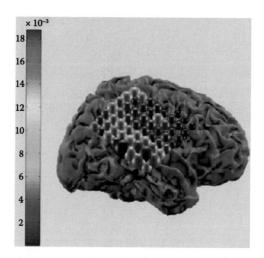

FIGURE 8.11 Granger causality flow between the cortical electrodes during an epileptic seizure.

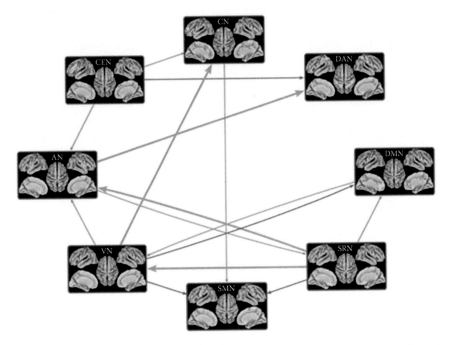

FIGURE 8.12 Granger causal relationships, grouped by their magnitude (strong influences in orange, medium influences in purple, and weak influences in green) between the resting state networks from the concatenated resting state fMRI data of 30 patients.

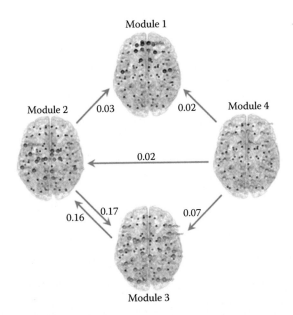

FIGURE 8.14 The causalities between the four modules of the fMRI application.

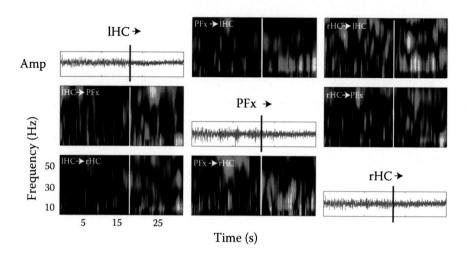

FIGURE 9.12 rPDC analysis applied to murine EEG data recorded bilaterally above the hippocampus (left: lHC, right: rHC), and prefrontal cortex (Pfx). Raw data are depicted on the diagonal (amplitude (Amp) in arbitrary units) over time in seconds. The lines indicate the time point of transition between NREM to REM sleep. For the rPDC spectra, direction of information flow is from column to row. Color coding (gray scale) indicates the intensity of coherence.

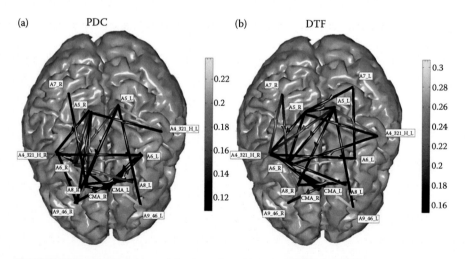

FIGURE 10.2 Cortical connectivity patterns obtained by PDC (a) and DTF (b) for the period preceding the subject's response during congruent trials in the beta (12–29 Hz) frequency band. The patterns are shown on the realistic head model and cortical envelope of the subject, obtained from sequential MRIs. The brain is seen from above, left hemisphere represented on the right side. The colors and sizes of the arrows code the level of strengths of the cortical connectivity observed among ROIs. The lighter and the bigger the arrows, the stronger the connections. Only the cortical connections statistically significant at $p < 0.01$ are reported.

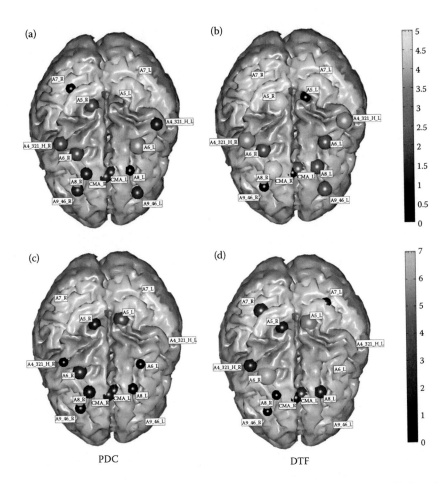

PDC   DTF

FIGURE 10.3 The inflow (a) and (b) and the outflow (c) and (d) patterns obtained for the beta frequency band from each ROI during the congruent trials. The brain is seen from above, left hemisphere represented on the right side. Different ROIs are depicted in colors (selected regions: left and right primary motor areas (BA4); posterior supplementary motor areas (pSMA: BA6); right and left parietal areas including BA7 and 5; Bas 8 and 9/46; left and right cingulate motor areas (CMA). (a) and (b): in red hues, the behavior of an ROI as the target of information flow from other ROIs. The larger the sphere, the higher the value of inflow or outflow for any given ROI. (c) and (d): in blue hues, the outflow of information from a single ROI toward all the others.

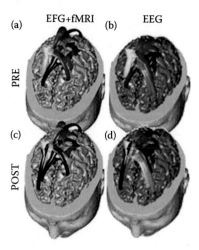

FIGURE 10.4 (a) and (b): cortical connectivity patterns estimated before the EMG onset (PRE) in the finger-tapping condition, while (c) and (d) are related to the connectivity patterns computed after the EMG onset. Connectivity was estimated by DTF. (a) and (c): DTF estimated on the cortical activity reconstructed by means of the neuro-electric information (EEG only). (b) and (d): cortical connectivity patterns estimated with the use of multimodal integration of neuroelectric and hemodynamic data (EEG+fMRI). Same conventions of Figure 10.2.

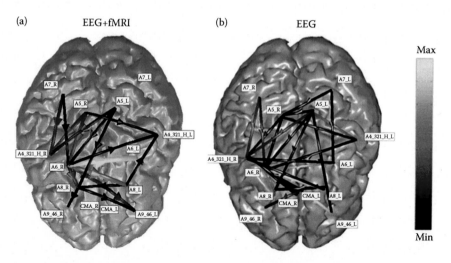

FIGURE 10.5 Presents two cortical connectivity patterns obtained with the DTF estimator, in the beta band, during the congruent condition in the Stroop task. (a): cortical connectivity pattern is estimated with the use of the EEG and fMRI information; (b): the connectivity pattern is estimated from EEG information only.

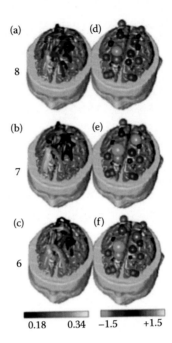

0.18   0.34   −1.5   +1.5

FIGURE 10.6 Panels from (a) through (c): Connectivity patterns related to the different model orders used for the MVAR of the finger tapping data, starting from (c) the order 6 to (a) the order 8. Panels from (d) through (f): Inflow (red hues) and outflow (blue hues) patterns related to the same range of MVAR model orders. The computed connectivity patterns show a substantial similarity across different values of the model order in an interval around the optimal value of 7 .

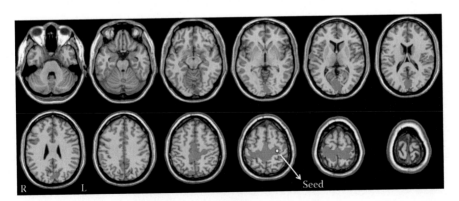

FIGURE 11.1 Correlation map for resting state data set ($r > 0.65$). The seed was placed at M1 in the left hemisphere (Medical image format inverts left and right, as can be seen by R and L above). Red indicates positive correlation. Note that the right hemisphere M1 and SMA are correlated to the seed voxel.

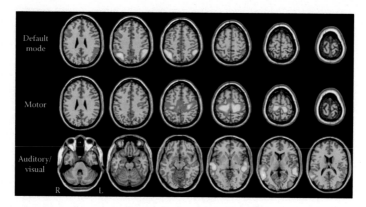

FIGURE 11.2 ICA mapping for resting state data set, illustrating components 3 (default mode network, top row), 11 (motor network, middle row), and 17 (auditory/visual network, bottom row) ($|z| > 2.00$). Red and blue indicates positive and negative coefficients, respectively.

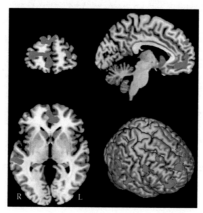

FIGURE 11.3 PPI analysis mapping for Stroop's task data set, with the seed placed at the ventral ACC ($|z| > 3.00$). Red indicates increased connectivity as an effect of increasing task difficulty (incongruent>congruent).

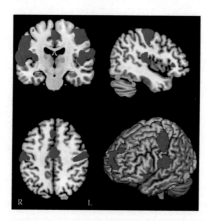

FIGURE 11.5 Granger causality maps for the resting state data, with the seed placed at the left hemisphere M1 ($|z| > 7.00$). The voxels which send information to the seed are depicted in blue, whereas those that receive information from the seed are depicted in green.

TABLE 7.1
Defining Variables According to PDC Type in
Equations 7.20 and 7.21

Variable	PDC	gPDC	ιPDC
$\mathcal{S}_n$	$\mathcal{I}_{2K} \otimes \mathcal{I}_K$	$\mathcal{I}_{2K} \otimes (\mathcal{I}_K \odot \Sigma_\mathbf{w})^{-1}$	$\mathcal{I}_{2K} \otimes (\mathcal{I}_K \odot \Sigma_\mathbf{w})^{-1}$
$\mathcal{S}_d$	$\mathcal{I}_{2K} \otimes \mathcal{I}_K$	$\mathcal{I}_{2K} \otimes (\mathcal{I}_K \odot \Sigma_\mathbf{w})^{-1}$	$\mathcal{I}_{2K} \otimes \Sigma_\mathbf{w}^{-1}$

2. $\mathbf{I}_j$, which is nonzero only for entries whose indices are of the form (l, m):
$(j-1)K + 1 \leq l = m \leq jK$.

As a result, PDC brands amount to quadratic form ratios in terms of $\mathbf{a}(\lambda)$. Rewriting PDC in this way facilitates in taking its derivatives with respect to the parameters as required by the Delta method, whose first implication in the current context is that PDC point estimate variance γ^2 at frequency λ (omitted ahead to simply notation) is given by

$$\gamma^2 = \mathbf{g_a} \Omega_\mathbf{a} \mathbf{g_a}^T + \mathbf{g_\sigma} \Omega_\sigma \mathbf{g_\sigma}^T, \tag{7.22}$$

for the respective gradient vectors

$$\mathbf{g_a} = 2\frac{\mathbf{a}^T \mathbf{I}_{ij}^c \mathcal{S}_n \mathbf{I}_{ij}^c}{\mathbf{a}^T \mathbf{I}_j^c \mathcal{S}_d \mathbf{I}_j^c \mathbf{a}} - 2\frac{\mathbf{a}^T \mathbf{I}_{ij}^c \mathcal{S}_n \mathbf{I}_{ij}^c \mathbf{a}}{(\mathbf{a}^T \mathbf{I}_j^c \mathcal{S}_d \mathbf{I}_j^c \mathbf{a})^2} \mathbf{a}^T \mathbf{I}_j^c \mathcal{S}_d \mathbf{I}_j^c \tag{7.23}$$

and

$$\mathbf{g_\sigma} = \frac{1}{\mathbf{a}^T \mathbf{I}_j^c \mathcal{S}_d \mathbf{I}_j^c \mathbf{a}} \left[(\mathbf{I}_{ij}^c \mathbf{a})^T \otimes (\mathbf{a}^T \mathbf{I}_{ij}^c) \right] \Theta_K \xi_n$$

$$- \frac{\mathbf{a}^T \mathbf{I}_{ij}^c \mathcal{S}_n \mathbf{I}_{ij}^c \mathbf{a}}{(\mathbf{a}^T \mathbf{I}_j^c \mathcal{S}_d \mathbf{I}_j^c \mathbf{a})^2} \left[(\mathbf{I}_j^c \mathbf{a})^T \otimes (\mathbf{a}^T \mathbf{I}_j^c) \right] \Theta_K \xi_d, \tag{7.24}$$

where the values of ξ_n and ξ_d are listed in Table 7.2 and where $\Theta_K = (\mathbf{T}_{2K,K} \otimes \mathbf{I}_{2K^2})(\mathbf{I}_K \otimes \text{vec}(\mathbf{I}_{2K} \otimes \mathbf{I}_K))$ with $\mathbf{T}_{L,M}$ standing for the commutation matrix (Lütkepohl, 2005).

When $\Sigma_\mathbf{w}$ is known *a priori* or does not need to be estimated, $\mathbf{g_\sigma} \Omega_\sigma \mathbf{g_\sigma}^T$ in Equation 7.22 is zero.

When $\mathbf{I}_{ij}^c \mathbf{a} = 0$ as for when the null hypothesis (7.3) holds, the latter gradients are nullified and the asymptotic result (7.2) no longer holds. This calls for the computation of the next term in the Delta method expansion and leads to Equation 7.4 where

TABLE 7.2
Defining Variables According to PDC Type in Equation 7.24

Variable	PDC	gPDC	ιPDC
ξ_n	0	$-\mathrm{diag}\left(\mathrm{vec}\left(\boldsymbol{S}_n^{-2}\right)\right)$	$-\mathrm{diag}\left(\mathrm{vec}\left(\boldsymbol{S}_n^{-2}\right)\right)$
ξ_d	0	$-\mathrm{diag}\left(\mathrm{vec}\left(\boldsymbol{S}_n^{-2}\right)\right)$	$-\boldsymbol{\Sigma}_{\mathbf{w}}^{-T} \otimes \boldsymbol{\Sigma}_{\mathbf{w}}^{-1}$

$l_k(\lambda)$ are the eigenvalues of $\mathcal{D} = \mathbf{L}^T \mathbf{I}_{ij}^c \boldsymbol{S}_n^{-1} \mathbf{I}_{ij}^c \mathbf{L}$, with $\mathbf{L}$ standing for the Choleski factor of $\bar{\boldsymbol{\Omega}}_{\mathbf{a}}$ with $q = \mathrm{rank}(\mathcal{D}) \leq 2$, which reduces to 1 whenever $\lambda \in \{0, \pm 0.5\}$ or $p = 1$ for all $\lambda \in [-0.5, 0.5]$.

REFERENCES

Baccalá, L. A., K. Sameshima, G. Ballester et al. 1998. Studying the interaction between brain structures via directed coherence and Granger causality. *Appl. Sign. Process.* 5: 40–48.

Baccalá, L. A. and K. Sameshima. 2001. Partial directed coherence: Some estimation issues. In *World Congress on Neuroinformatics*, 546–553, Vienna, Austria.

Baccalá, L. A., D. Y. Takahashi, and K. Sameshima. 2006. Computer intensive testing for the influence between time series. In R. Schelter, M. Winterhalter, and J. Timmer (Eds.), *Handbook of Time Series Analysis*, 411–436. Berlin: Wiley-VCH.

Baccalá, L. A., D. Y. Takahashi, and K. Sameshima. 2007. Generalized partial directed coherence. In *15th International Conference on Digital Signal Processing* (DSP2007), 163–166, Cardiff, Wales, UK.

Baccalá, L. A., C. S. N. de Brito, D. Y. Takahashi et al. 2013. Unified asymptotic theory for all partial directed coherence forms. *Philos. Trans. R. Soc. A* 371: 1–13.

Clarke, A. J. 2008. *An Introduction to the Dynamics of El Nino & the Southern Oscillation*. Oxford: Academic Press.

de Brito, C., L. A. Baccalá, D. Y. Takahashi et al. 2010. Asymptotic behavior of generalized partial directed coherence. In *Engineering in Medicine and Biology Society (EMBC), 2010 Annual International Conference of the IEEE*, 1718–1721, Buenos Aires, Argentina.

Geweke, J. 1982. Measurement of linear dependence and feedback between multiple time series. *J. Am. Stat. Assoc.* 77: 304–324.

Geweke, J., R. Meese, and W. Dent. 1983. Comparative tests of causality in temporal systems. *J. Econometrics 21*: 161–194.

Geweke, J. 1984. Inference and causality in economic time series. In Z. Griliches and M. D. Intrilligator (Eds.), *Handbook of Econometrics*, Vol. 11, 1101–1144. Amsterdam: Elsevier.

Imhof, J. P. 1961. Computing the distribution of quadratic forms in normal variables. *Biometrika 48*: 419–426.

Kamiński, M., M. Ding, W. Truccolo et al. 2001. Evaluating causal relations in neural systems: Granger causality, directed transfer function and statistical assessment of significance. *Biol. Cybern.* 85: 145–157.

Lütkepohl, H. 1993. *Introduction to Multiple Time Series Analysis* (2nd ed.). Berlin: Springer.

Lütkepohl, H. 2005. *New Introduction to Multiple Time Series Analysis*. New York: Springer.

Patnaik, P. B. 1949. The non-central χ^2 and F-distributions and their applications. *Biometrika* *36*: 202–232.

Prichard, D. and J. Theiler. 1994. Generating surrogate data for time-series with several simultaneously measured variables. *Phys. Rev. Lett.* *73*: 951–954.

Sarachik, E. S. and M. A. Cane. 2010. *The El Niño-Southern Oscillation Phenomenon.* Cambridge: Cambridge University Press.

Schelter, B., M. Winterhalder, M. Eichler et al. 2005. Testing for directed influences among neural signals using partial directed coherence. *J. Neurosci. Methods* *152*: 210–219.

Takahashi, D. Y., L. A. Baccalá, and K. Sameshima. 2007. Connectivity inference between neural structures via partial directed coherence. *J. Appl. Stat.* *34*: 1259–1273.

Takahashi, D., L. Baccalá, and K. Sameshima. 2008. Partial directed coherence asymptotics for VAR processes of infinite order. *Int. J. Bioelectromag.* *10*: 31–36.

Takahashi, D. Y., L. A. Baccalá, and K. Sameshima. 2010. Information theoretic interpretation of frequency domain connectivity measures. *Biol. Cybern.* *103*: 463–469.

van der Vaart, A. W. 1998. *Asymptotic Statistics.* Cambridge: Cambridge University Press.

Wang, J.-Y., H.-T. Zhang, J.-Y. Chang et al. 2008. Anticipation of pain enhances the nociceptive transmission and functional connectivity within pain network in rats. *Mol. Pain* *4*: 34.

Section II

Extensions

8 Nonlinear Parametric Granger Causality in Dynamical Networks

Daniele Marinazzo, Wei Liao, Mario Pellicoro, and Sebastiano Stramaglia

CONTENTS

8.1 INTRODUCTION

Determining how the brain is connected is a crucial point in neuroscience. Advances in imaging techniques guarantee an immediate improvement in our knowledge of structural connectivity. A constant computational and modeling effort has to be done in order to optimize and adapt functional and effective connectivity to the qualitative and quantitative changes in data and physiological applications. We are interested in the paths of information flow throughout the brain and what this can tell us about the functionality of a healthy and pathological brain. Every time we record brain activity, we can imagine that we are monitoring the activity at the nodes of a network. This activity is dynamical and sometimes chaotic. Dynamical networks (Barabási, 2002) model physical and biological behavior in many applications.

Dynamic network synchronization is influenced by its topology (Boccaletti et al., 2006). There is a great need for valid methods for inferring network structure from time-series data; a dynamic version of Bayesian Networks has been proposed in Ghahramani (1998). A method for detecting the topology of dynamical networks, based on chaotic synchronization, has been proposed in Yu et al. (2006); a recent approach deals with the case of a low number of samples and proposed methods rooted on L1 minimization (Napoletani and Sauer, 2008).

In this chapter, we propose a method to infer the structure of the network from the dynamics that we record at its nodes.

Granger causality has become the method of choice to determine whether and how two time series exert causal influences on each other (Hlaváčková-Schindler et al., 2007). This approach is based on prediction: if the prediction error of the first time series is reduced by including measurements from the second one in the linear regression model, then the second time series is said to have a causal influence on the first one. This frame has been used in many fields of science, including neural systems (Kamiński et al., 2001), rheo-chaos (Ganapathy et al., 2007), and cardio-vascular variability (Faes et al., 2008). The estimation of linear Granger causality from Fourier and wavelet transforms of time-series data has been addressed recently (Dhamala et al., 2008).

Kernel algorithms work by embedding data into a Hilbert space and searching for linear relations in that space (Vapnik, 1998). The embedding is performed implicitly by specifying the inner product between pairs of elements (Shawe-Taylor and Cristianini, 2004). We have recently exploited the properties of kernels to provide nonlinear measures of bivariate Granger causality (Marinazzo et al., 2008b). We reformulated linear Granger causality and introduced a new statistical procedure to handle overfitting (Paluš et al., 1993) in the linear case. Our new formulation was then generalized to the nonlinear case by means of the kernel trick (Shawe-Taylor and Cristianini, 2004), thus obtaining a method with the following two main features: (i) the nonlinearity of the regression model can be controlled by choosing the kernel function, and (ii) the problem of false-causalities, which arises as the complexity of the model increases, is addressed by a selection strategy of the eigenvectors of a reduced Gram matrix whose range represents the additional features due to the second time series.

In this chapter, we describe in detail the kernel Granger approach and address the use of Granger causality to estimate, from multivariate time-series data, the topology and the drive–response relationships of a dynamical network. To this aim, we generalize our method in Marinazzo et al. (2008a) to the case of multivariate data.

This chapter is organized as follows. In the next section, we describe bivariate Granger causality, whereas the multivariate generalization is described in Section 8.3. Some methodological issues are discussed in Section 8.4. In Section 8.5, we describe some applications, and in Section 8.6 we focus on the relationships between Granger causality and redundancy. After a discussion, the conclusions are summarized in Section 8.7.

8.2   BIVARIATE GRANGER CAUSALITY

In this section, we review the kernel method for Granger causality proposed in Marinazzo et al. (2008b). Let us start with the linear case.

8.2.1   LINEAR GRANGER CAUSALITY

To understand the nonlinear approaches to Granger causality as generalizations of the linear case, we briefly review the linear case. We begin by a brief review of the standard linear approach to Granger causality. Suppose that we express the temporal dynamics of a stationary time series $\{\xi_n\}_{n=1,\dots,N+m}$ as an autoregressive model of order m:

$$\xi_n = \sum_{j=1}^{m} A_j \, \xi_{n-j} + E_n$$

and by a bivariate autoregressive model which also takes into account a simultaneously recorded time series $\{\eta_n\}_{n=1,\dots,N+m}$:

$$\xi_n = \sum_{j=1}^{m} A'_j \, \xi_{n-j} + \sum_{j=1}^{m} B_j \, \eta_{n-j} + E'_n.$$

The coefficients of the models are calculated by standard least-squares optimization; there exist several criteria for the optimal choice of the model order m; see the discussion in Section 8.4.

The concept of Granger causality is (Hlaváčková-Schindler et al., 2007): η Granger causes ξ if the variance of residuals E' is significantly smaller than the variance of residuals E, as happens when coefficients B_j are jointly significantly different from zero. This can be tested by performing an F-test or Levene's test for equality of variances (Geweke, 1982). An index measuring the strength of the causal interaction is then defined as

$$\delta = 1 - \frac{\langle E'^2 \rangle}{\langle E^2 \rangle}, \tag{8.1}$$

where $\langle \cdot \rangle$ means averaging over n (note that $\langle E \rangle = \langle E' \rangle = 0$). Exchanging the roles of the two time series, one may equally test causality in the opposite direction, that is, to check whether ξ Granger causes η.

This basic definition can be extended to nonlinear models and to the multivariate case. Let us start with the bivariate nonlinear extension.

In order to deal with the possible nonlinearities in the measured signals, different approaches to nonlinear Granger causality have been proposed. This concern originated first in the field of econometrics (Bell et al., 1996; Hiemstra and Jones, 1994; Teräsvirta, 1996; Warne, 2000). The issue was then transferred to neuroscience in 1999, with a seminal paper by Freiwald et al. (1999), where the extension to the nonlinear case was performed by specifying a linear autoregressive model whose coefficients depended on the previous states of the system. In that paper, the authors pointed out that in order to be suitable to evaluate causality, a nonlinear model should be matched to the variable and unpredictable characteristics of the system. The issue of detecting causal influences in neural (and thus most probably nonlinear) systems is discussed in Valdes et al. (1999). A different approach is discussed in Chávez et al. (2003): the nonlinear Granger causality is seen there in probabilistic terms as a violation of the Markov property, which states that the probability of finding a system in a given state in the future, given its past, does not change if we condition it to the past of another independent variable (Schreiber, 2000).

A similar method is discussed in Gourévitch et al. (2006): the violation of the Markov property is tested there by means of the correlation integral (Grassberger and Procaccia, 1983).

A straightforward extension of Equation 8.1 to evaluate nonlinear Granger causality is proposed by Bezruchko et al. (2008), in which the autoregression model is constructed in the form of a polynomial of order p. As correctly pointed out by the authors, indeed, for short time series, the prevision tends to be unstable when increasing the dimensionality of the model and the order of the polynomial. In the same paper, the authors propose to evaluate nonlinear connectivity looking at the phases, albeit not in the framework of Granger causality.

All those methods are unfortunately particularly amenable to overfitting of the learning scheme.

The approach described in this chapter is rooted in machine learning and regression theory. We start by redefining linear, bivariate Granger causality in the framework of information theory. Then, we will use the properties of regularized kernel Hilbert spaces (RKHS) to extend the definition to the nonlinear case.

Granger causality is based on the idea that a predictive model for a given time series can be improved by including information from other time series. Modeling a given phenomenon ultimately means learning something from the available observations and then using this information to make predictions on the future of the system. This approach is named *regression* in machine learning. Finding such patterns in the observations can also be used to classify the data into several groups.

We use the shorthand notations:

$$X_i = (\xi_i, \dots, \xi_{i+m-1})^\top,$$

$$Y_i = (\eta_i, \dots, \eta_{i+m-1})^\top$$

and $x_i = \xi_{i+m}$, for $i = 1, \dots, N$. We treat these quantities as N realizations of the stochastic variables X, Y, and x. Let us denote $\mathbf{X}$ the $m \times N$ matrix having vectors X_i as columns, and $\mathbf{Z}$ the $2m \times N$ matrix having vectors $Z_i = (X_i^\top, Y_i^\top)^\top$ as columns. The values of x are organized in a vector $\mathbf{x} = (x_1, \dots, x_N)^\top$. In full generality, we assume that each component of X and Y has a zero mean, and that vector $\mathbf{x}$ has a zero mean and is normalized, that is, $\mathbf{x}^\top \mathbf{x} = 1$.

Now, for each $i = 1, \dots, N$, we define

$$\tilde{x}_i = \sum_{j=1}^{m} A_j \, \xi_{i+m-j},$$

$$\tilde{x}_i' = \sum_{j=1}^{m} A_j' \, \xi_{i+m-j} + \sum_{j=1}^{m} B_j \, \eta_{i+m-j}.$$

In the two cases, the vectors $\tilde{\mathbf{x}} = (\tilde{x}_1, \ldots, \tilde{x}_N)^\top$ and $\tilde{\mathbf{x}}' = (\tilde{x}_1', \ldots, \tilde{x}_N')^\top$ are the estimated values by linear regression. It is easy to show that $\tilde{\mathbf{x}}$ and $\tilde{\mathbf{x}}'$ have the following geometrical interpretation. Let $H \subseteq \Re^N$ be the range of the $N \times N$ matrix $\mathbf{K} = \mathbf{X}^\top \mathbf{X}$. Then $\tilde{\mathbf{x}}$ is the projection of $\mathbf{x}$ on H. In other words, calling $\mathbf{v}_1, \ldots, \mathbf{v}_m$ the (orthonormal) eigenvectors of $\mathbf{K}$ with nonvanishing eigenvalue and

$$P = \sum_{i=1}^{m} \mathbf{v_i} \mathbf{v_i}^\top$$

the projector on the space H, we have $\tilde{\mathbf{x}} = P\mathbf{x}$. Let us define $\mathbf{y} = \mathbf{x} - P\mathbf{x}$. Analogously, $\tilde{\mathbf{x}}' = P'\mathbf{x}$, P' being the projector on the $2m$-dimensional space $H' \subseteq \Re^N$, equal to the range of the matrix $\mathbf{K}' = \mathbf{Z}^\top \mathbf{Z}$. Moreover, it is easy to show that

$$\delta = \frac{\tilde{\mathbf{x}}'^\top \tilde{\mathbf{x}}' - \tilde{\mathbf{x}}^\top \tilde{\mathbf{x}}}{1 - \tilde{\mathbf{x}}^\top \tilde{\mathbf{x}}}. \tag{8.2}$$

Now note that $H \subseteq H'$. Therefore, we may decompose H' as follows: $H' = H \oplus H^\perp$, where $H^\perp$ is the space of all vectors of H' orthogonal to all vectors of H. $H^\perp$ corresponds to the additional features due to the inclusion of $\{\eta\}$ variables. Calling $P^\perp$ the projector on $H^\perp$, we can write

$$\delta = \frac{\|P^\perp \mathbf{y}\|^2}{1 - \tilde{\mathbf{x}}^\top \tilde{\mathbf{x}}}. \tag{8.3}$$

Now, we note that $H^\perp$ is the range of the matrix

$$\tilde{\mathbf{K}} = \mathbf{K}' - \mathbf{K}'P - P\left(\mathbf{K}' - \mathbf{K}'P\right) = \mathbf{K}' - P\mathbf{K}' - \mathbf{K}'P + P\mathbf{K}'P.$$

Indeed, for any $\mathbf{u} \in \Re^N$, we have $\tilde{\mathbf{K}}\mathbf{u} = \mathbf{v} - P\mathbf{v}$, where $\mathbf{v} = \mathbf{K}'(\mathbf{I} - P)\mathbf{u} \in H'$, and $\tilde{\mathbf{K}}\mathbf{u} \in H^\perp$. It follows that $H^\perp$ is spanned by the set of the eigenvectors, with nonvanishing eigenvalue, of $\tilde{\mathbf{K}}$. Calling $\mathbf{t}_1, \ldots, \mathbf{t}_m$ these eigenvectors, we have

$$\delta = \sum_{i=1}^{m} r_i^2, \tag{8.4}$$

where r_i is Pearson's correlation coefficient of $\mathbf{y}$ and $\mathbf{t_i}$ (since the overall sign of $\mathbf{t_i}$ is arbitrary, we can assume that r_i is positive). Let π_i be the probability that r_i is, due to chance, obtained by Student's t test. Since we are dealing with multiple comparison, we use the Bonferroni correction to select the eigenvectors $\mathbf{t}_{i'}$, correlated with $\mathbf{y}$, with the expected fraction of false-positive q (equal to 0.05). Therefore, we calculate a new

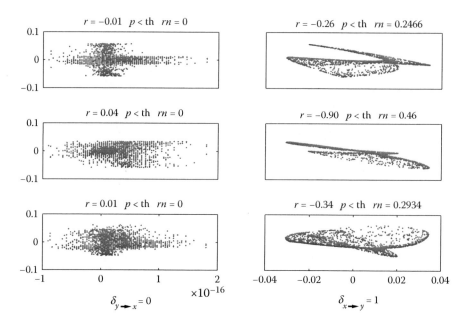

FIGURE 8.1 The eigenvalues of the Gram matrix described in the methods are plotted versus the realizations of the variables in the case of two unidirectionally coupled chaotic maps for the three components.

causality index by summing, in Equation 8.4, only over the $\{r_{i'}\}$ such that $\pi_{i'} < q/m$, thus obtaining a *filtered* linear Granger causality index:

$$\delta_F = \sum_{i'} r_{i'}^2. \tag{8.5}$$

It is assumed that δ_F measures the causality $\eta \to \xi$, without further statistical testing. Nonetheless, for data sets with multiple comparisons and multiple patients, a null distribution can be generated using a surrogate time series. An example of this procedure in the case of unilateral influence between two coupled chaotic maps as described below is reported in Figure 8.1.

8.2.2 KERNEL GRANGER CAUSALITY

In this subsection, we describe the generalization of linear Granger causality to the nonlinear case, using methods from the theory of RKHS (Shawe-Taylor and Cristianini, 2004).

Kernel methods provide an efficient way to perform nonlinear regression. In this case, linear regression is performed in a more suitable space, named the feature space, built from a nonlinear combination of the input variables. Figure 8.2 shows a cartoon example of a function which maps the features corresponding to two different sets in a suitable space where the separator is a line. In this way, regression is nonlinear

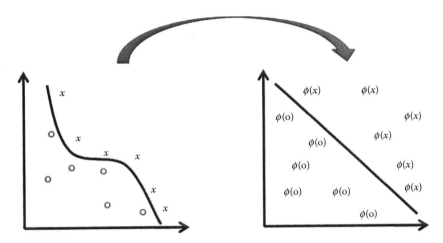

FIGURE 8.2   Cartoon sketch of a given function ϕ which maps the features in a space in which they are linearly separated.

with respect to the input variables, still keeping all the optimal parameters of the linear case. The feature representation can be obtained with nonlinear polynomials or other nonlinear functions. Another key advantage of kernel methods is the kernel trick (Vapnik, 1998): the input space does not need to be explicitly mapped to a high-dimensional feature space by means of a nonlinear function $\Phi : x \to \Phi(x)$. Instead, kernel-based methods take advantage of the fact that most linear methods only require the computation of dot products. Hence, the trick in kernel-based learning is to substitute an appropriate kernel function in the input space for the dot products in the feature space, that is,

$$k : \mathfrak{R}^N \times \mathfrak{R}^N \to \mathfrak{R} \tag{8.6}$$

such that $k(x, x') = \phi(x)^\top \phi(x')$. More precisely, kernel-based methods are restricted to the particular class of kernel functions k that correspond to dot products in feature spaces and hence only implicitly map the input space to the corresponding feature space (Lemm et al., 2011).

Given a kernel function K, with spectral representation $K(X, X') = \sum_a \lambda_a \phi_a(X) \phi_a(X')$ (see Mercer's theorem; Vapnik, 1998), we consider H, the range of the $N \times N$ Gram matrix $\mathbf{K}$ with elements $K(X_i, X_j)$. In order to make the mean of all variables $\phi_a(X)$ zero, we replace $\mathbf{K} \to \mathbf{K} - P_0\mathbf{K} - \mathbf{K}P_0 + P_0\mathbf{K}P_0$, where P_0 is the projector onto the one-dimensional subspace spanned by the vector such that each component is equal to unity (Shawe-Taylor and Cristianini, 2004); in the following, we assume that this operation has been performed on each Gram matrix. As in the linear case, we calculate $\tilde{\mathbf{x}}$, the projection of $\mathbf{x}$ onto H. Owing to the spectral representation of K, $\tilde{\mathbf{x}}$ coincides with the linear regression of $\mathbf{x}$ in the feature space spanned by $\sqrt{\lambda_a}\phi_a$, the eigenfunctions of K; the regression is nonlinear in the original variables.

While using both X and Y to predict x, we evaluate the Gram matrix $\mathbf{K}'$ with elements $K'_{ij} = K(Z_i, Z_j)$. The regression values now form the vector $\tilde{\mathbf{x}}'$ equal to the projection of $\mathbf{x}$ on H', the range of $\mathbf{K}'$. Before evaluating the filtered causality index, as in the linear case, we note that not all kernels may be used to evaluate Granger causality. Indeed, if Y is statistically independent of X and x, then $\tilde{x}'$ and $\tilde{x}$ should coincide in the limit $N \to \infty$. This property, invariance of the risk minimizer when statistically independent variables are added to the set of input variables, is satisfied only by suitable kernels, as discussed in Ancona and Stramaglia (2006). In the following, we consider two possible choices, which fulfill the invariance requirement.

Inhomogeneous polynomial kernel. The inhomogeneous polynomial (IP) kernel of integer order p is

$$K_p(X, X') = \left(1 + X^\top X'\right)^p.$$

In this case, the eigenfunctions are made of all the monomials, in the input variables, up to the p-th degree. The dimension of the space H is $m_1 = 1/B(p+1, m+1) - 1$, where B is the beta function, and $p = 1$ corresponds to the linear regression. The dimension of space H' is $m_2 = 1/B(p+1, 2m+1) - 1$. As in the linear case, we note that $H \subseteq H'$ and decompose $H' = H \oplus H^\perp$. Subsequently, we calculate $\tilde{\mathbf{K}} = \mathbf{K}' - \mathbf{P}\mathbf{K}' - \mathbf{K}'\mathbf{P} + \mathbf{P}\mathbf{K}'\mathbf{P}$; the dimension of the range of $\tilde{\mathbf{K}}$ is $m_3 = m_2 - m_1$. Along the same lines as those described in the linear case, we construct the kernel Granger causality taking into account only the eigenvectors of $\tilde{\mathbf{K}}$, which pass the Bonferroni test:

$$\delta_F^K = \sum_{i'} r_{i'}^2, \tag{8.7}$$

the sum being only over the eigenvectors of $\tilde{\mathbf{K}}$ with probability $\pi_{i'} < q/m_3$.

Gaussian kernel. The Gaussian kernel reads:

$$K_\sigma(X, X') = \exp\left(-\frac{(X - X')^\top (X - X')}{2\sigma^2}\right), \tag{8.8}$$

and depends on the width σ. σ controls the complexity of the model: the dimension of the range of the Gram matrix decreases as σ increases. As in previous cases, we may consider H, the range of the Gram matrix $\mathbf{K}$, and H', the range of $\mathbf{K}'$, but in this case the condition $H \subseteq H'$ would not necessarily hold; therefore, some differences in the approach must be undertaken. We call L the m_1-dimensional span of the eigenvectors of $\mathbf{K}$ whose eigenvalue is not smaller than $\mu\lambda_{max}$, where λ_{max} is the largest eigenvalue of $\mathbf{K}$ and μ is a small number (we use 10^{-6}). We evaluate $\tilde{x} = Px$, where P is the projector on L. After evaluating the Gram matrix $\mathbf{K}'$, the following matrix is considered:

$$\mathbf{K}^* = \sum_{i=1}^{m_2} \rho_i \mathbf{w}_i \mathbf{w}_i^\top, \tag{8.9}$$

where $\{\mathbf{w}\}$ are the eigenvectors of $\mathbf{K}'$, and the sum is over the eigenvalues $\{\rho_i\}$ not smaller than μ times the largest eigenvalue of $\mathbf{K}'$. Then we evaluate $\tilde{\mathbf{K}} = \mathbf{K}^* - \mathbf{P}\mathbf{K}^* - \mathbf{K}^*\mathbf{P} + \mathbf{P}\mathbf{K}^*\mathbf{P}$, and denote $P^\perp$ the projector onto the m_3-dimensional range of $\tilde{\mathbf{K}}$. Note that the condition $m_2 = m_1 + m_3$ may not be strictly satisfied in this case (however, in our experiments, we find that the violation of this relation is always very small, if any). The kernel Granger causality index for the Gaussian kernel is then constructed as in the previous case; see Equation 8.7.

8.2.3 Connection with Information-Theoretic Approaches

Granger causality is deeply related to information-theoretic approaches such as the transfer entropy in Schreiber (2000). Let $\{\xi_n\}_{n=1,\ldots,N+m}$ be a time series that may be approximated by a stationary Markov process of order m, that is, $p(\xi_n|\xi_{n-1},\ldots,\xi_{n-m}) = p(\xi_n|\xi_{n-1},\ldots,\xi_{n-m-1})$. We will use the shorthand notation $X_i = (\xi_i,\ldots,\xi_{i+m-1})^\top$ and $x_i = \xi_{i+m}$, for $i = 1,\ldots,N$, and treat these quantities as N realizations of the stochastic variables X and x. The minimizer of the risk functional

$$\mathcal{R}[f] = \int dX\, dx\, (x - f(X))^2\, p(X,x) \tag{8.10}$$

represents the best estimate of x, given X, and corresponds (Papoulis, 1985) to the regression function $f^*(X) = \int dx p(x|X)x$. Now, let $\{\eta_n\}_{n=1,\ldots,N+m}$ be another time series of simultaneously acquired quantities, and denote $Y_i = (\eta_i,\ldots,\eta_{i+m-1})^\top$. The best estimate of x, given X and Y, is now: $g^*(X,Y) = \int dx p(x|X,Y)x$. If the generalized Markov property holds, that is, $p(x|X,Y) = p(x|X)$, then $f^*(X) = g^*(X,Y)$ and the knowledge of Y does not improve the prediction of x. Transfer entropy (Schreiber, 2000) is a measure of the violation of the property mentioned above: it follows that Granger causality implies nonzero transfer entropy. An important bridge between autoregressive and information-theoretic approaches to data-driven causal inference is the fact that Granger causality and transfer entropy are entirely equivalent for Gaussian variables, as has been demonstrated in Barnett et al. (2009). Owing to the finiteness of N, the risk functional cannot be evaluated; we consider the empirical risk

$$ER[f] = \sum_{i=1}^{N} (x_i - f(X_i))^2,$$

and the search for the minimum of *ER* is constrained in a suitable functional space, called hypothesis space; the simplest choice is the space of all linear functions, corresponding to linear regression. In the following, we propose a geometrical description of linear Granger causality. For each $\alpha \in \{1,\ldots,m\}$, the samples of the α-th component of X form a vector $\mathbf{u}_\alpha \in \mathfrak{R}^N$; without loss of generality, we assume that each $\mathbf{u}_\alpha$ has a zero mean and that $\mathbf{x} = (x_1,\ldots,x_N)^\top$ is normalized and has a zero mean. We denote $\tilde{x}_i$ by the value of the linear regression of x versus X, evaluated at X_i. The vector $\tilde{\mathbf{x}} = (\tilde{x}_1,\ldots,\tilde{x}_N)^\top$ can be obtained as follows. Let $H \subseteq \mathfrak{R}^N$ be the span of $\mathbf{u}_1,\ldots,\mathbf{u}_m$; then $\tilde{\mathbf{x}}$ is the projection of $\mathbf{x}$ on H. In other words, calling P the projector on the space H, we have $\tilde{\mathbf{x}} = P\mathbf{x}$. Moreover, the prediction error, given X, is $\epsilon_x = ||\mathbf{x} - \tilde{\mathbf{x}}||^2 = 1 - \tilde{\mathbf{x}}^\top\tilde{\mathbf{x}}$. Calling **X** the $m \times N$ matrix having vectors $\mathbf{u}_\alpha$ as rows, H coincides with the range of the $N \times N$ matrix $\mathbf{K} = \mathbf{X}^\top\mathbf{X}$.

8.3 MULTIVARIATE KERNEL CAUSALITY

Let's consider the set $\{\xi(a)_n\}_{n=1,\ldots,N+m}$ of M simultaneously recorded series of length $N + m$. In order to put in evidence the drive–response pattern in this system, one may evaluate the bivariate Granger causality, described in the previous sections, between every pair of time series. It is recommended, however, to treat the data set as a whole, thus generalizing kernel Granger causality to the multivariate case as follows. We denote

$$X(c)_i = (\xi(c)_i, \ldots, \xi(c)_{i+m-1})^\top,$$

for $c = 1, \ldots, M$, and $i = 1, \ldots, N$. In order to evaluate the causality $\{\xi(a)\} \to \{\xi(b)\}$, we define, for $i = 1, \ldots, N$,

$$Z_i = (X(1)_i^\top, \ldots, X(a)_i^\top, \ldots, X(M)_i^\top)^\top,$$

containing all the input variables, and

$$X_i = (X(1)_i^\top, \ldots, X(M)_i^\top)^\top,$$

containing all the input variables but those related to $\{\xi(a)\}$. Gram matrices $\mathbf{K}$ and $\mathbf{K}'$ are then evaluated: $K_{ij} = K(X_i, X_j)$ and $K'_{ij} = K(Z_i, Z_j)$. The target vector is now $\mathbf{x} = (\xi(b)_{1+m}, \ldots, \xi(b)_{N+m})^\top$. Along the same lines as in the bivariate case, for IP kernel or the Gaussian one, we then calculate the causality index as in Equation 8.6: it is denoted by $\delta_F^K (a \to b)$ and measures the strength of causality $a \to b$, taking into account all the available variables. Repeating these steps for all a and b, the causality pattern in the data set is evaluated. Note that the threshold for Bonferroni's correction, in the multivariate case, must be lowered by the number of pairs $M(M - 1)/2$.

8.4 METHODOLOGICAL ISSUES

8.4.1 CHOICE OF THE MODEL ORDER

The order of the autoregressive model can be chosen according to the Akaike criterion or the Bayesian Information criterion (BIC), or by cross-validation (Lemm et al., 2011). Given the information theoretical approach of KGC, we think that LOO-based cross-validation (see Lemm et al., 2011, for details) is a more suitable choice. According to this approach, the optimal model order is the one which minimizes the leave-one-out error. In Figures 8.3 and 8.4, we report the empirical risk and the leave-one-out error for a sample of the ECoG and fMRI data used in the applications, respectively.

8.4.2 CHOICE OF THE WIDTH OF THE GAUSSIAN KERNEL

In the case of kernel causality by means of a Gaussian kernel, the choice of the width of the Gaussian, σ, may be done, for example, by cross-validation. Empirically, we find that this yields a value typically ranging in [5,10] if normalized data are under consideration. Moreover, it is also interesting to look at the change in the causality

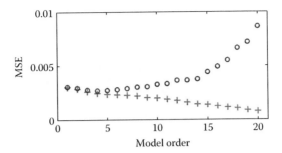

FIGURE 8.3 Leave-one-out cross-validation for the choice of the model order. Empirical risk (crosses) and LOO error (empty circles) are represented versus the order of the model for the regression of one channel of the electrocorticogram versus another.

pattern as the width σ is varied: as σ controls the complexity of the model, this variation gives insight into the degree of nonlinearity of the flow of information in the system at hand.

8.4.3 STABILITY OF THE MODEL

Despite being stabler compared to other methods, KGC becomes less precise and eventually fails when the number of data points becomes too small, when the number of variables becomes too large, when the order of the model is too high, or when there is a combination of these factors. One should keep in mind that, to obtain one stochastic realization of the data, a number of points not smaller than the number of variables is needed. Of course, only one stochastic realization is not enough to get any

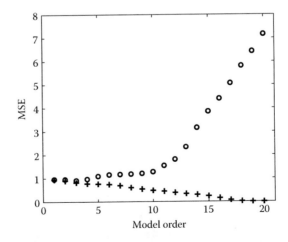

FIGURE 8.4 Leave-one-out cross-validation for the choice of the model order. Empirical risk (crosses) and LOO error (empty circles) are represented versus the order of the model for the regression of one ROI of one fMRI versus another.

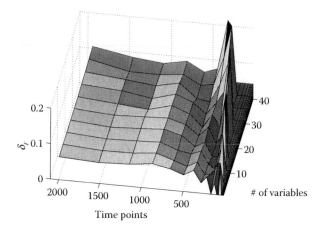

FIGURE 8.5 (**See color insert.**) Granger-causal relationships of a given pair of variables in a nonlinear AR model for a different number of variables and time-series length.

statistics, so we need a reasonably elevated number of these realizations (the rows of the Z matrix). To explore empirically the limits of the algorithm, one could start from a low model order and a high number of sample points and then gradually reduce the number of points or increase the order of the model. When the algorithm starts to fail, one observes a rapid increase of the causality value, followed by a drop. In a way, this is analogous to what happens for the coupled maps when the coupling is too strong, that is, no more information flow is detected. A typical situation for a given interaction in an autoregressive nonlinear model versus added independent variables and length of the time series is reported in Figure 8.5.

8.5 ANALYSIS OF DYNAMICAL NETWORKS

In this section, we show applications on simulated and real data sets.

8.5.1 TWO COUPLED MAPS

First, we consider two unidirectionally coupled noisy logistic maps:

$$
\begin{aligned}
x_{n+1} &= 1 - ax_n^2 + s\eta_n, \\
y_{n+1} &= (1-e)(1-ay_n^2) + e(1-ax_n^2) + s\xi_n;
\end{aligned} \tag{8.11}
$$

$\{\eta\}$ and $\{\xi\}$ are unit variance Gaussianly distributed noise terms (the parameter s determines their relevance), $a = 1.8$ and $e \in [0, 1]$ represents the coupling $x \to y$. In the noise-free case ($s = 0$), a transition to complete synchronization (Pikowski et al., 2001) occurs at $e = 0.37$. Varying e and s, we have considered runs of N iterations, after a transient of 10^3, and evaluated δ_F, using the kernel with $p = 2$, in both directions. We find that $\delta_F(Y \to X)$ is zero for all values of e, s, and N. On the other hand, $\delta_F(X \to Y)$ is zero at e smaller than a threshold e_c; see Figure 8.6. $\delta_F(X \to Y)$ is zero

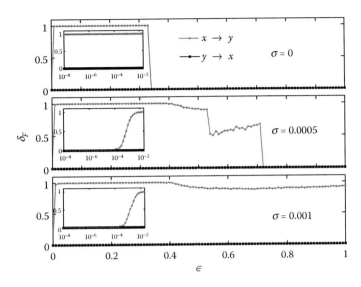

FIGURE 8.6 The Granger causality index between the two coupled chaotic maps x and y versus the coupling strength ϵ for three noise amplitudes. The insets report the same quantities for small values of the coupling.

also at complete synchronization, as there is no information transfer in this regime. These results are generated with the code **KG_two_ccm.m** included in the CD.

8.5.2 Network of Chaotic Maps

Let us consider a coupled map lattice of n nodes, with equations, for $i = 1, \ldots, n$:

$$x_{i,t} = \left(1 - \sum_{j=1}^{n} c_{ij}\right)\left(1 - ax_{i,t-1}^2\right)$$
$$+ \sum_{j=1}^{n} c_{ij}\left(1 - ax_{j,t-1}^2\right) + s\tau_{i,t}, \tag{8.12}$$

where a, s, and τ's are the same as in Equation 8.16, and c_{ij} represents the coupling $j \to i$. We fix $n = 34$ and construct couplings as follows. We consider the well-known Zachary data set (Zachary, 1977), an undirected network of 34 nodes. We assign a direction to each link, with equal probability, and set c_{ij} equal to 0.05, for each link of the directed graph thus obtained, and zero otherwise. The network is displayed in Figure 8.7: here the goal is to estimate this directed network from the measurements of time series on nodes.

The multivariate Granger analysis, described in the previous section, perfectly recovers the underlying network using the IP kernel with $p = 2$, $m = 1$, and $N = 2000$. The code to perform this analysis is in the script **example_zachary_34_nodes.m**.

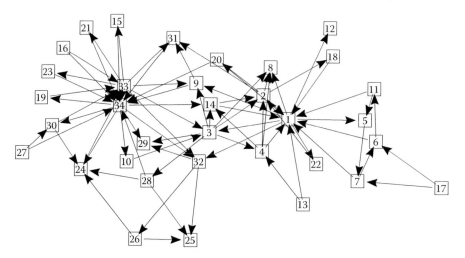

FIGURE 8.7   The directed network of 34 nodes obtained by assigning randomly a direction to links of the Zachary network.

Note that while evaluating $\delta_F^K(j \to i)$, for all i and j, 39,270 Pearson's coefficients r are calculated. Their distribution is represented in Figure 8.8: there is a strong peak at $r = 0$ (corresponding to projections that are discarded), and a very low number of projections with a rather large value of r, up to about $r = 0.6$. It is interesting to describe the results in terms of a threshold for correlations. Given a threshold value r,

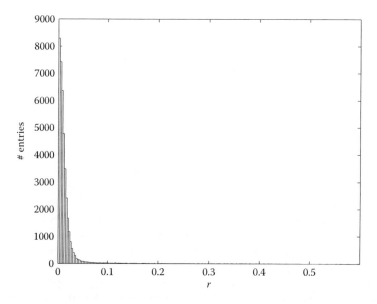

FIGURE 8.8   The distribution of the 39,270 r-values calculated while evaluating the causality indexes of the coupled map lattice (see the text).

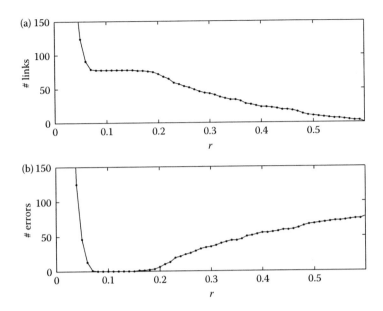

FIGURE 8.9   (a) Concerning the coupled map lattice, the horizontal axis represents the threshold for the values of r; the plot shows the number of links of the directed network constructed from the projections whose Pearson's coefficient exceeds the threshold. (b) The total number of errors, in the reconstructed network, is plotted versus the threshold r. At large r, the errors are due only to missing links, whereas at large r the errors are due only to links that do not exist in the true network and are recovered.

we select the correlation coefficients whose value is greater than r. We then calculate the corresponding causality indexes $\delta_F^K(j \to i)$, and construct the directed network whose links correspond to nonvanishing elements of $\delta_F^K(j \to i)$. In Figure 8.9a, we plot the total number of links of the reconstructed network as a function of the threshold r: the curve shows a plateau, around $r = 0.1$, corresponding to a directed network which is stable against variations of r. At the plateau, 428 projections are selected, which coincide with those selected by means of Bonferroni's test. In Figure 8.9b, we plot the number of errors (the sum of the number of links that exist in the true network and are not recovered, plus the number of links that exist only in the recovered network) versus the threshold r: the plateau leads to perfect reconstruction of the original network.

We stress that a large number of samples is needed to recover the underlying network: in the typical case, we find that the network is perfectly reconstructed if $N \geq 2\,000$, whereas if N is further lowered, some errors in the reconstructed network appear. Moreover, it is important to observe that, although all couplings c_{ij} have the same magnitude, the causality strengths $\delta_F^K(j \to i)$ depend on the link, as is shown in Figure 8.10. Granger causality is a measure of the information being transferred from one time series to another, and it should not be identified with couplings.

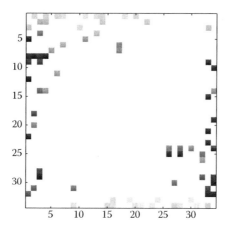

FIGURE 8.10 The Granger causality indexes $\delta_F^K (j \rightarrow i)$, for all pairs of maps, are represented in a gray scale. White square means zero causality, whereas black squares correspond to the maximal causality observed (0.55).

8.5.3 LOCALIZATION OF EPILEPTICAL FOCI IN AN ELECTROCORTICOGRAM

Connectivity analysis can prove very useful in determining the onset of an epileptic seizure. We analyzed the intracranial seizure recording obtained from a pediatric patient with medically intractable neocortical-onset epilepsy. This data set is contained in the eConnectome software (He et al., 2011) and is part of a larger data set described in Wilke et al. (2010). It contains recordings from silastic electrode grids (interelectrode distance of 1 cm) implanted on the cortical surface. See Wilke et al.

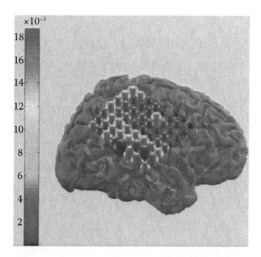

FIGURE 8.11 **(See color insert.)** Granger causality flow between the cortical electrodes during an epileptic seizure.

(2010) for a complete description. Our analysis with $p = 2$ and $m = 4$ revealed a putative focus in a specific region of the cortex (Figure 8.11).

8.5.4 INFLUENCES BETWEEN RESTING STATE NETWORKS IN fMRI

It has been recently proposed that brain activity at rest is spatially organized in a set of specific coherent patterns, called resting state networks (RSNs). These patterns reflect the functional architecture of specific areas, which are commonly modulated during active behavioral tasks. In this analysis, we consider eight RSNs: default-mode, dorsal attention, core, central-executive, self-referential, somatosensory, visual, and auditory. We considered 30 subjects from Zang's group resting state data from 1000 functional connectomes. For each subject, a total of 225 volumes (TR = 1, 32 slices) were acquired. The first 25 volumes were discarded to ensure steady-state longitudinal magnetization. The RNSs were individuated by independent component analysis (ICA), as described in Liao et al. (2009).

The spatial maps of the eight RSNs obtained with this analysis are illustrated in Figure 8.12. On the basis of our classification results, and those of a large number of RSN studies, the eight ICs associated with RSNs can be described as follows (see Liao et al., 2009, for a complete description):

- RSN 1: A network of regions typically referred to as the DMN. This network has been suggested to be involved in episodic memory and the self-projection. It involves the posterior cingulate cortex (PCC)/precuneus region, bilateral inferior parietal gyrus, angular gyrus, middle temporal gyrus, superior frontal gyrus, and medial frontal gyrus.

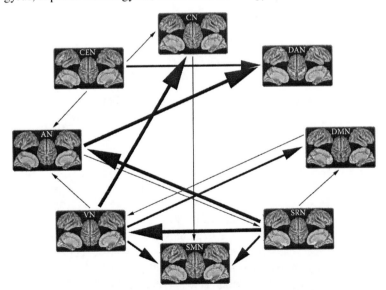

FIGURE 8.12 (**See color insert.**) Granger causal relationships, grouped by their magnitude (strong influences in orange, medium influences in purple, and weak influences in green) between the resting state networks from the concatenated resting state fMRI data of 30 patients.

- RSN 2: A network overlapping with the DAN, which is thought to mediate goal-directed top-down processing. The network, which is left-lateralized, primarily involves the middle and superior occipital gyrus, the parietal gyrus, inferior and superior parietal gyrus, and the middle and superior frontal gyrus.
- RSN 3: A right-lateralized network putatively associated with central-executive network (CEN), whose key regions include the dorsal lateral prefrontal cortices and the posterior parietal cortices.
- RSN 4: A network dedicated to visual processing (VN), which includes the inferior, middle, and superior occipital gyrus, the temporal–occipital regions, along with the superior parietal gyrus.
- RSN 5: A network that primary encompasses the bilateral, middle, and superior temporal gyrus, the Heschl gyrus, the insular cortex, the temporal pole, and corresponds to the auditory system (AN).
- RSN 6: A network corresponding to the sensory-motor function (SMN). This network includes the pre- and postcentral gyrus, the primary sensorimotor cortices, and the supplementary motor area.
- RSN 7: A network putatively related to SRN. It includes the ventromedial prefrontal cortex (vMPFC), the medial orbital prefrontal cortex (MOPFC), the gyrus rectus, and the pregenual anterior cingulate gyrus (PACC).
- RSN 8: A network associated with task control function, namely, core network (CN) whose key regions include the anterior cingulate, the bilateral insular, and the dorso-lateral prefrontal cortices.

The RSN time series from 30 patients were concatenated, and the Kernel Granger Causality algorithm was applied with $p = 2$ and $m = 1$. The results are reported in Figure 8.12.

Repeating the same analysis on 30 different subjects, we found qualitatively the same pattern of influences.

8.6   REDUNDANCY AND GRANGER CAUSALITY

An important notion in information theory is the redundancy in a group of variables, formalized in Paluš et al. (1993) as a generalization of the mutual information. A formalism to recognize redundant and synergetic variables in neuronal ensembles has been proposed in Schneidman et al. (2003); the information theoretic treatments of groups of correlated degrees of freedom can reveal their functional roles in complex systems. In Angelini et al. (2010), it has been shown that the presence of redundant variables influences the performance of multivariate Granger causality and an approach to exploit redundancy, so as to identify functional patterns in data, has been proposed. To describe the approach, a quantitative definition to recognize redundancy and synergy in the frame of causality is in order. We will show that maximization of the total causality is connected to the detection of groups of redundant variables.

Let us consider n time series $\{x_\alpha(t)\}_{\alpha=1,\ldots,n}$; the previous state vectors are denoted

$$X_\alpha(t) = (x_\alpha(t-m), \ldots, x_\alpha(t-1)),$$

m being the window length (the choice of m can be made using the standard cross-validation scheme). Let $\epsilon(x_\alpha|\mathbf{X})$ be the mean-squared error prediction of x_α. The straightforward generalization of Granger causality for sets of variables is

$$\delta(B \to A) = \sum_{\alpha \in A} \frac{\epsilon(x_\alpha|\mathbf{X}\backslash B) - \epsilon(x_\alpha|\mathbf{X})}{\epsilon(x_\alpha|\mathbf{X}\backslash B)}, \tag{8.13}$$

where A and B are two disjoint subsets of $\{1,\ldots,n\}$, and $\mathbf{X}\backslash B$ means the set of all variables except for those X_β with $\beta \in B$. On the other hand, the unnormalized version of it, that is,

$$\delta^u(B \to A) = \sum_{\alpha \in A} \{\epsilon(x_\alpha|\mathbf{X}\backslash B) - \epsilon(x_\alpha|\mathbf{X})\}, \tag{8.14}$$

can easily be shown to satisfy the following interesting property: if $\{X_\beta\}_{\beta \in B}$ are statistically independent and their contributions in the model for A are additive, then

$$\delta^u(B \to A) = \sum_{\beta \in B} \delta^u(\beta \to A). \tag{8.15}$$

In order to identify the informational character of a set of variables B, concerning the causal relationship $B \to A$, we remind that, in general, synergy occurs if B contributes to A with more information than the sum of all its variables, whereas redundancy corresponds to situations with the same information being shared by the variables in B. We make quantitative these notions and define the variables in B *redundant* if $\delta^u(B \to A) > \sum_{\beta \in B} \delta^u(\beta \to A)$, and *synergetic* if $\delta^u(B \to A) < \sum_{\beta \in B} \delta^u(\beta \to A)$. In order to justify these definitions, first we observe that the case of independent variables (and additive contributions) does not fall either in the redundancy case or in the synergetic case, due to Equation 8.15, as it should be. Moreover, we describe the following example for two variables X_1 and X_2. If X_1 and X_2 are redundant, then removing X_1 from the input variables of the regression model does not have a great effect, as X_2 provides the same information as X_1; this implies that $\delta^u(X_1 \to A)$ is nearly zero. The same reasoning holds for X_2; hence, we expect that $\delta^u(\{X_1, X_2\} \to A) > \delta^u(X_1 \to A) + \delta^u(X_2 \to A)$. Conversely, let us suppose that X_1 and X_2 are synergetic, that is, they provide some information about A only when both the variables are used in the regression model; in this case, $\delta^u(\{X_1, X_2\} \to A)$, $\delta^u(X_1 \to A)$, and $\delta^u(X_2 \to A)$ are almost equal, and therefore $\delta^u(\{X_1, X_2\} \to A) < \delta^u(X_1 \to A) + \delta^u(X_2 \to A)$.

The following analytically tractable example is useful to clarify these notions. Consider two stationary and Gaussian time series $x(t)$ and $y(t)$ with $\langle x^2(t) \rangle = \langle y^2(t) \rangle = 1$ and $\langle x(t)y(t) \rangle = C$; they correspond, for example, to the asymptotic regime of the autoregressive system

$$x_{t+1} = ax_t + by_t + \sigma \xi_{t+1}^{(1)}$$
$$\tag{8.16}$$
$$y_{t+1} = bx_t + ay_t + \sigma \xi_{t+1}^{(2)},$$

where ξ are i.i.d. unit variance Gaussian variables, $\mathcal{C} = 2ab/(1 - a^2 - b^2)$ and $\sigma^2 = 1 - a^2 - b^2 - 2ab\mathcal{C}$. Considering the time series $z_{t+1} = A(x_t + y_t) + \sigma'\xi_{t+1}^{(3)}$ with $\sigma' = \sqrt{1 - 2A^2(1 + \mathcal{C})}$, we obtain for $m = 1$:

$$\delta^u(\{x, y\} \to z) - \delta^u(x \to z) - \delta^u(y \to z) = A^2(\mathcal{C} + \mathcal{C}^2). \qquad (8.17)$$

Hence, x and y are redundant (synergetic) for z if $\mathcal{C}$ is positive (negative). Turning to consider $w_{t+1} = B\, x_t \cdot y_t + \sigma''\xi_{t+1}^{(4)}$ with $\sigma'' = \sqrt{1 - B^2(1 + 2\mathcal{C})^2}$, and using the polynomial kernel with $p = 2$, we have

$$\delta^u(\{x, y\} \to z) - \delta^u(x \to z) - \delta^u(y \to z) = B^2(4\mathcal{C}^2 - 1); \qquad (8.18)$$

x and y are synergetic (redundant) for w if $|\mathcal{C}| < 0.5$ ($|\mathcal{C}| > 0.5$).

The presence of redundant variables leads to an under-estimation of their causality when the standard multivariate approach is applied (as is clear from the discussion above, this is not the case for synergetic variables). Redundant variables should be grouped to get a reliable measure of causality, and to characterize interactions in a more compact way. As is clear from the discussion above, grouping of redundant variables is connected to maximization of the unnormalized causality index (8.14) and, in the general setting, can be made as follows. For a given target α_0, we call B the set of the remaining $n - 1$ variables. The partition $\{A_\ell\}$ of B, maximizing the total causality

$$\Delta = \sum_\ell \delta^u(A_\ell \to x_{\alpha_0}),$$

consists of groups of redundant variables with respect to (w.r.t.) the target α_0. The global information character of the system can be summarized by searching for the partition $\{A_\ell\}$ of the system's variables maximizing the sum of the causalities between every pair of subsets

$$\Gamma = \sum_\ell \sum_{\ell' \neq \ell} \delta^u(A_\ell \to A_{\ell'}).$$

However, the search over all the partitions is unfeasible except for small systems. A simplified to find group of causally redundant variables (groups of variables sharing the same information about the future of the system), which can also be applied to large systems, reads as follows (Marinazzo et al., 2010). We assume that the essential features of the dynamics of the system under consideration are captured using just a small number of characteristic modes. One can then use principal components analysis to obtain a compressed representation of the future state of the system. A pairwise measure of the redundancy, w.r.t. the prediction of the next configuration of the modes, is evaluated, thus obtaining a weighted graph. Finally, by maximizing the modularity (Newman, 2006b), one finds the natural modules of this weighted graph and identifies them with the groups of redundant variables. Now, we go into the details of the method.

Step 1 Denoting $X_i(t) = x_i(t - 1)$ as the lagged times series, the modes of the systems may be extracted by principal components analysis as follows. Calling **x**

the $n \times T$ matrix with elements $x_i(t)$, we denote $\{u_\alpha(t)\}_{\alpha=1,...,n_\lambda}$ the (normalized) eigenvectors of the matrix $\mathbf{x}^\top \mathbf{x}$ corresponding to the largest n_λ eigenvalues. The T-dimensional vectors $u_\alpha(t)$ summarize the dynamics of the system; the correlations among variables at different times determine to what extent the modes u may be predicted on the basis of the $X_i(t)$ variables.

Step 2 The variables which are significatively correlated with the modes u are selected. For each i and each α, we evaluate the probability $p_{i\alpha}$ that the correlation between X_i and u_α is, due to chance, obtained by Student's t test. We compare $p_{i\alpha}$ with the 5% confidence level after Bonferroni's correction (the threshold is $0.05/(n \times n_\lambda)$) and retain only those variables X_i, which are significantly correlated with at least one mode. The variables thus selected will be denoted $\{Y_i(t)\}_{i=1,...,N}$, N being their cardinality.

Step 3 The next step of the approach is the introduction of a bivariate measure of redundancy as follows. For each pair of variables Y_i and Y_j, we denote P_i the projector onto the one-dimensional space spanned by Y_i and P_j the projector onto the space corresponding to Y_j; P_{ij} is the projector onto the bi-dimensional space spanned by Y_i and Y_j. Then, we define:

$$c_{ij} = \sum_{\alpha=1}^{n_\lambda} \left(||P_i u_\alpha||^2 + ||P_j u_\alpha||^2 - ||P_{ij} u_\alpha||^2 \right); \qquad (8.19)$$

c_{ij} is positive (negative) if the variables Y_i and Y_j are redundant (synergetic) w.r.t. the prediction of the future of the system. In other words, if Y_i and Y_j share the same information about u, then c_{ij} is positive.

Step 4 Finally, the matrix c_{ij} is used to construct a weighted graph of N nodes, the weight of each link measuring the degree of redundancy between the two variables connected by that link. By maximization of the modularity, the number of modules, as well as their content, is extracted from the weighted graph. Each module is recognized as a group of variables sharing the same information about the future of the system. As an example, we consider fMRI data from a subject in resting conditions, with sampling frequency 1 Hz, and the number of samples equal to 500. A prior brain atlas is utilized to parcellate the brain into 90 cortical and subcortical regions, and a single time series is associated with each region. All the 90 time series are then bandpassed in the range 0.01–0.08 Hz., so as to reduce the effects of low-frequency drift and high-frequency noise.

In Figure 8.13, top-left, we depict N, the number of selected variables, as a function of n_λ. For $n_\lambda > 3$, all the 90 regions are recognized as influencing the future of the system. For each value of n_λ, the quantities c_{ij} are evaluated. In Figure 8.13, top-left, we depict N, the number of variables selected as relevant for the future of the system, as a function of n_λ. For $n_\lambda > 3$, all the 90 regions are selected. As a following step, the Granger causality values c_{ij} for each pair of variables are evaluated at each value of n_λ. The modularity algorithm described in Newman (2006a) is then applied to the matrix of Granger causality values to retrieve the number of modules M_m in which it is optimally divided. In Figure 8.13, top-right, this number is depicted against n_λ;

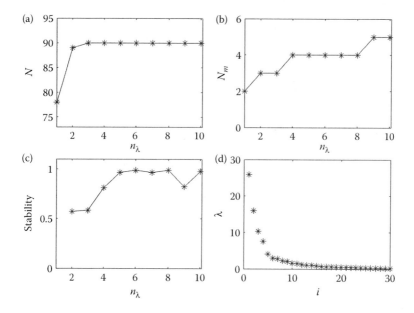

FIGURE 8.13 (a) Concerning the fMRI application, the number of selected regions N is plotted versus n_λ, the number of modes. (b) The number of modules, obtained by modularity maximization, of the matrix c_{ij}, whose elements measure the pairwise redundancy. (c) The measure of the stability of the partition, going from $n_\lambda - 1$ to n_λ, is plotted versus n_λ. (d) The eigenvalues of the matrix $x^\top x$ are depicted.

this plot suggests the presence of four modules for $4 < n_\lambda < 8$. These values are the stablest, as it is clear from Figure 8.13, bottom-left, where we plot the measure of the stability of the partition while going from $n_\lambda - 1$ to n_λ. It follows that the optimal value of n_λ is four, corresponding to a graph structure with four modules and modularity equal to 0.3. We find that module 1 includes brain regions from the ventral medial frontal cortices, which are primarily specialized for the anterior default mode network. This brain network is also termed as a self-referential network, which has been assumed to filter, select, and provide those stimuli which are relevant for the self of a particular person. Module 2 is typically referred to as the posterior default mode network, which exhibits high levels of activity during the resting state and decreases the activity for processes of internal-oriented mental activity, such as mind wandering, episodic memory, and environmental monitoring. Module 3 mainly corresponds to the executive control network, assumed to be dedicated to adaptive task control. Module 4 refers to some brain regions in the occipital lobe that are primarily specialized for visual processing and the others in the subcortical cortices. In Figure 8.13, bottom-right, the eigenvalues of the matrix $x^\top x$ are depicted. These results confirm that the brain is highly redundant, as most information theoretic work shows (see, e.g., Narayanan et al., 2005). Averaging the time series belonging to each module, we obtain four time series and evaluate the causalities between them: the result is displayed in Figure 8.14. It is interesting to observe that module 4 influences all the three

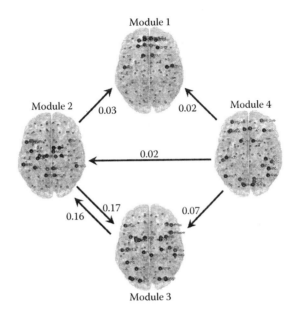

FIGURE 8.14   (**See color insert.**) The causalities between the four modules of the fMRI application.

other modules but is not influenced by them, and is an out-degree hub; this is consistent with the fact that it corresponds to the subcortical brain. Another striking feature is the clear interdependencies between modules 2 and 3. The generalization of this approach to a nonlinear measure of redundancy is straightforward, as it corresponds to the use of the unnormalized kernel Granger causality.

8.7   DISCUSSION

Our method of analysis of dynamical networks is based on a recent measure of Granger causality between time series, rooted on kernel methods, whose magnitude depends on the amount of flow of information from one series to another. By the definition of Granger causality, our method allows an analysis of networks containing cycles. First, we have demonstrated the effectiveness of the method on a network of chaotic maps with links obtained by assigning a direction to the edges of the well-known Zachary data set, using a nonlinear kernel: perfect reconstruction of the directed network is achieved provided that a sufficient number of samples is available.

For a given application, one should choose the proper kernel out of the many possible classes (Shawe-Taylor and Cristianini, 2004).

Detecting cause and effect influences between components of a complex system is an active multidisciplinary area of research in these years. The kernel approach presented here provides a statistically robust tool to assess drive–response relationships in many fields of science.

REFERENCES

Ancona, N. and S. Stramaglia. 2006. An invariance property of predictors in kernel-induced hypothesis spaces. *Neural Comput. 18*: 749–759.

Angelini, L., M. de Tommaso, D. Marinazzo et al. 2010. Redundant variables and Granger causality. *Phys. Rev. E 81*: 037201.

Barabási, A.-L. 2002. *Linked*. Cambridge: Perseus Publishing.

Barnett, L., A. B. Barrett, and A. K. Seth. 2009. Granger causality and transfer entropy are equivalent for Gaussian variables. *Phys. Rev. Lett. 103*: 238701.

Bell, D., J. Kay, and J. Malley. 1996. A non-parametric approach to non-linear causality testing. *Econ. Lett. 51*: 7–18.

Bezruchko, B. P., V. I. Ponomarenko, M. D. Prokhorov et al. 2008. Modeling nonlinear oscillatory systems and diagnostics of coupling between them using chaotic time series analysis: Applications in neurophysiology. *Physics-Uspekhi 51*: 304.

Boccaletti, S., D.-H. Hwang, M. Chávez et al. 2006. Synchronization in dynamical networks: Evolution along commutative graphs. *Phys. Rev. E 74*: 016102.

Chávez, M., J. Martinerie, and M. Le Van Quyen. 2003. Statistical assessment of nonlinear causality: Application to epileptic EEG signals. *J. Neurosci. Methods 124*: 113–128.

Dhamala, M., G. Rangarajan, and M. Ding. 2008. Estimating Granger causality from Fourier and wavelet transforms of time series data. *Phys. Rev. Lett. 100*: 018701.

Faes, L., G. Nollo, and K. H. Chon. 2008. Assessment of Granger causality by nonlinear model identification: Application to short-term cardiovascular variability. *Ann. Biomed. Eng. 36*: 381–395.

Freiwald, W. A., P. Valdes, J. Bosch et al. 1999. Testing non-linearity and directedness of interactions between neural groups in the macaque inferotemporal cortex. *J. Neurosci. Methods 94*: 105–119.

Ganapathy, R., G. Rangarajan, and A. K. Sood. 2007. Granger causality and cross recurrence plots in rheochaos. *Phys. Rev. E 75*: 016211.

Geweke, J. 1982. Measurement of linear dependence and feedback between multiple time series. *J. Am. Stat. Assoc. 77*: 304–313.

Ghahramani, Z. 1998. Learning dynamic Bayesian networks. *Lect. Notes Comput. Sc. 1387*: 168–197.

Gourévitch, B., R. L. Bouquin-Jeannès, and G. Faucon. 2006. Linear and nonlinear causality between signals: Methods, examples and neurophysiological applications. *Biol. Cybern. 95*: 349–369.

Grassberger, P. and I. Procaccia. 1983. Measuring the strangeness of strange attractors. *Physica D: Nonlinear Phenomena 9*: 189–208.

He, B., Y. Dai, L. Astolfi et al. 2011. eConnectome: A MATLAB toolbox for mapping and imaging of brain functional connectivity. *J. Neurosci. Methods 195*: 261–269.

Hiemstra, C. and J. D. Jones. 1994. Testing for linear and nonlinear Granger causality in the stock price–volume relation. *J. Finan. 49*: 1639–1664.

Hlaváčková-Schindler, K., M. Paluš, M. Vejmelka et al. 2007. Causality detection based on information-theoretic approaches in time series analysis. *Phys. Rep. 441*: 1–46.

Kamiński, M., M. Ding, W. A. Truccolo et al. 2001. Evaluating causal relations in neural systems: Granger causality, directed transfer function and statistical assessment of significance. *Biol. Cybern. 85*: 145–157.

Lemm, S., B. Blankertz, T. Dickhaus et al. 2011. Introduction to machine learning for brain imaging. *NeuroImage 56*: 387–399.

Liao, W., D. Mantini, Z. Zhang et al. 2009. Evaluating the effective connectivity of resting state networks using conditional Granger causality. *Biol. Cybern. 102*: 57–69.

Marinazzo, D., M. Pellicoro, and S. Stramaglia. 2008a. Kernel-Granger causality and the analysis of dynamical networks. *Phys. Rev. E 77*: 056215.

Marinazzo, D., M. Pellicoro, and S. Stramaglia. 2008b. Kernel method for nonlinear Granger causality. *Phys. Rev. Lett. 100*, 144103.

Marinazzo, D., W. Liao, M. Pellicoro et al. 2010. Grouping time series by pairwise measures of redundancy. *Phys. Lett. A 374*: 4040–4044.

Napoletani, D. and T. D. Sauer. 2008. Reconstructing the topology of sparsely connected dynamical networks. *Phys. Rev. E 77*: 026103.

Narayanan, N. S., E. Y. Kimchi, and M. Laubach. 2005. Redundancy and synergy of neuronal ensembles in motor cortex. *J. Neurosci. 25*: 4207–4216.

Newman, M. E. J. 2006a. Finding community structure in networks using the eigenvectors of matrices. *Phys. Rev. E 74*, 036104.

Newman, M. E. J. 2006b. Modularity and community structure in networks. *Proc. Natl. Acad. Sci. USA 103*: 8577–8582.

Paluš, M., V. Albrecht, and I. Dvořák. 1993. Information theoretic test for nonlinearity in time series. *Phys. Lett. A 175*: 203–209.

Papoulis, A. 1985. *Probability, Random Variables, and Stochastic Processes*. New York: McGraw-Hill.

Pikowski, A., M. Rosemblum, and J. Kurths. 2001. *Synchronization, a Universal Concept in Nonlinear Sciences*. Cambridge: The Cambridge University Press.

Schneidman, E., W. Bialek, and M. J. Berry. 2003. Synergy, redundancy, and independence in population codes. *J. Neurosci. 23*: 11539–11553.

Schreiber, T. 2000. Measuring information transfer. *Phys. Rev. Lett. 85*: 461–464.

Shawe-Taylor, J. and N. Cristianini. 2004. *Kernel Methods for Pattern Analysis*. Cambridge: The Cambridge University Press.

Teräsvirta, T. 1996. Modelling economic relationships with smooth transition regressions. In *Handbook of Applied Economic Statistics*, ed. D. Giles, and A. Ullah, 507–552. Stockholm: The Economic Research Institute.

Valdes, P. A., J. C. Jimenez, J. Riera et al. 1999. Nonlinear EEG analysis based on a neural mass model. *Biol. Cybern. 81*: 415–424.

Vapnik, V. N. 1998. *The Nature of Statistical Learning Theory*. New York: Springer-Verlag.

Warne, A. 2000. Causality and regime inference in a Markow switching VAR. *Sveriges Riksbank, Stockholm 118*: 1–41.

Wilke, C., W. van Drongelen, M. Kohrman et al. 2010. Neocortical seizure foci localization by means of a directed transfer function method. *Epilepsia 51*: 564–572.

Yu, D., M. Righero, and L. Kocarev. 2006. Estimating topology of networks. *Phys. Rev. Lett. 97*: 188701.

Zachary, W. 1977. Information-flow model for conflict and fission in small groups. *J. Anthropol. Res. 33*: 452–473.

9 Time-Variant Estimation of Connectivity and Kalman's Filter

*Linda Sommerlade, Marco Thiel, Bettina Platt,
Andrea Plano, Gernot Riedel, Celso Grebogi,
Wolfgang Mader, Malenka Mader, Jens Timmer,
and Björn Schelter*

CONTENTS

9.1   INTRODUCTION

Interactions of network structures are of particular interest in many fields of research since they promise to disclose the underlying mechanisms. Additional information on the direction of interactions is important to identify causes and their effects as well as loops in a network. This knowledge may then be used to determine the best target for interference, for example, stimulation of a certain brain region, with the network.

Several analysis techniques exist to determine relationships and/or causal influences in multivariate systems (e.g., Kamiński and Blinowska, 1991; Schack et al., 1995; Kamiński et al., 1997; Korzeniewska et al., 1997; Arnold et al., 1998; Pompe et al., 1998; Dahlhaus, 2000; Eichler, 2000; Halliday and Rosenberg, 2000; Rosenblum and Pikovsky, 2001; Rosenblum et al., 2002; Dahlhaus and Eichler, 2003). Partial directed coherence was introduced to determine causal influences (Baccalá and Sameshima, 2001). It was successfully applied in neuroscience (Sameshima and Baccalá, 1999; Nicolelis and Fanselow, 2002). This frequency-domain measure extends the bivariate coherence analysis (Brockwell and Davis, 1998) and multivariate graphical models applying partial coherences (Brillinger, 1981; Dahlhaus, 2000). It has recently been generalized to renormalized partial directed coherence (Schelter et al., 2009), which also allows the interpretation of the results in terms of the strength of an interaction.

Renormalized partial directed coherence (rPDC) is based on modeling time series by vector autoregressive (VAR) processes. Commonly, coefficients of the autoregressive processes are estimated via the multivariate Yule–Walker equations (Lütkepohl, 1993). The estimated coefficients of the autoregressive process are then Fourier transformed, leading to rPDC, a quantity in the frequency domain.

Some links in a network may not be constant in time. The strengths of interactions may change or a link may exist only for certain periods of time. If the interactions are not constant, the process is nonstationary. Since the aforementioned estimation algorithm assumes stationarity, it cannot deal with time-dependent coefficients. Windowing the time series and performing the analysis in each block results in an approximation for nonstationary data, but rapid changes compared to the block length in parameters cannot be detected this way. Estimation methods designed for time-dependent autoregressive coefficients (Grenier, 1983; Kitagawa and Gersch, 1996; Wan and van der Merwe, 2000) rely on *a priori* information either about their functional form or at least about the smoothness of changes. A time-dependent implementation of autoregressive models and a corresponding parameter fitting algorithm are crucial in many applications. We provide this by combining the Kalman filter (Hamilton, 1994; Harvey, 1994; Shumway and Stoffer, 2000) with rPDC. This leads to a time-resolved analysis technique detecting causal influences between processes in multivariate systems.

This chapter is structured as follows. First, in Section 9.2, renormalized partial directed coherence (rPDC) is summarized. Second, time-dependent influences are discussed in Section 9.3. Third, a time-dependent parameter estimation technique is presented in Section 9.4. Examples of the combination of rPDC and the Kalman filter are illustrated in Section 9.5, followed by an application to sleep transitions in mice in Section 9.6.

9.2   RENORMALIZED PARTIAL DIRECTED COHERENCE

When analyzing the direction of influences in a network, a definition of causality is mandatory. A widely used definition is based on Granger's linear predictability criterion and thus called Granger-causality (Granger, 1969). It is based on the common sense conception that causes precede their effects in time. Granger-causality is directly linked to autoregressive processes (Granger, 1969; Geweke, 1982, 1984).

Based on Granger's definition of causality, Baccalá and Sameshima (1998) introduced partial directed coherence to examine causal influences in multivariate systems (Baccalá et al., 1998; Sameshima and Baccalá, 1999; Baccalá and Sameshima, 2001) . This frequency-domain measure was refined by renormalization to also allow conclusions about the strength of interactions (Schelter et al., 2009). In order to estimate partial directed coherence, a VAR process is fitted to the multivariate time series.

Therefore, an n-dimensional vector autoregressive model of order p (VAR[p])

$$\vec{x}(t) = \sum_{r=1}^{p} \mathbf{a}_r \vec{x}(t-r) + \boldsymbol{\varepsilon}_x(t) \qquad \boldsymbol{\varepsilon}_x(t) \sim \mathcal{N}(\vec{0}|\boldsymbol{\Sigma}) \tag{9.1}$$

is considered, where $\boldsymbol{\varepsilon}_x$ denotes independent Gaussian white noise with the covariance matrix $\boldsymbol{\Sigma}$ and $\mathbf{a}_r$ the coefficient matrices of the VAR. A single component of such a VAR process can be interpreted in physical terms as a combination of relaxators and damped oscillators (Honerkamp, 1993). The spectral matrix of a VAR process is given by (Brockwell and Davis, 1998)

$$\mathbf{S}(\omega) = \mathbf{H}(\omega)\boldsymbol{\Sigma}\,\mathbf{H}^H(\omega), \tag{9.2}$$

where the superscript $(\cdot)^H$ denotes the Hermitian transpose. The transfer matrix $\mathbf{H}(\omega)$ is defined as

$$\mathbf{H}(\omega) = \left[\mathbf{I} - \bar{A}(\omega)\right]^{-1} = [\mathbf{A}(\omega)]^{-1}, \tag{9.3}$$

where $\mathbf{I}$ is the n-dimensional identity matrix and $\bar{A}(\omega)$ is given by the Fourier transform of the coefficients

$$\bar{A}_{kj}(\omega) = \sum_{r=1}^{p} \mathbf{a}_{r,kj}\, e^{-i\omega r}. \tag{9.4}$$

Breaking the symmetry of $\mathbf{S}(\omega)$ in Equation 9.2 leads to partial directed coherence (Baccalá and Sameshima, 2001)

$$\pi_{ij}(\omega) = \frac{|\mathbf{A}_{ij}(\omega)|}{\sqrt{\sum_k |\mathbf{A}_{kj}|^2(\omega)}} \in [0,1]. \tag{9.5}$$

A point wise α-significance level for partial directed coherence has been derived (Schelter et al., 2005). The denominator for a given influence from x_j onto x_i depends on the influences of x_j onto all other processes. Thus, higher partial directed coherence values at a given frequency do not necessarily correspond to a stronger influence from x_j onto x_i, but could also be due to weaker influences from x_j onto all other processes at the frequency of interest.

To overcome this, consider the two-dimensional vector of Fourier-transformed coefficients

$$\mathbf{X}_{kj}(\omega) = \begin{pmatrix} \mathrm{Re}(\mathbf{A}_{kj}(\omega)) \\ \mathrm{Im}(\mathbf{A}_{kj}(\omega)) \end{pmatrix} \tag{9.6}$$

with $\mathbf{X}_{kj}(\omega)'\mathbf{X}_{kj}(\omega) = |\mathbf{A}_{kj}(\omega)|^2$. The corresponding estimator $\hat{\mathbf{X}}_{kj}(\omega)$ with $\hat{\mathbf{A}}_{kj}(\omega)$ substituted for $\mathbf{A}_{kj}(\omega)$ is asymptotically normally distributed with the mean $\mathbf{X}_{kj}(\omega)$ and the covariance matrix $\mathbf{V}_{kj}(\omega)/N$ (Schelter et al., 2009), where N is the number of data points and

$$\mathbf{V}_{kj}(\omega) = \sum_{l,m=1}^{p} \mathbf{R}_{jj}^{-1}(l,m) \mathbf{\Sigma}_{kk} \begin{pmatrix} \cos(l\omega)\cos(m\omega) & \cos(l\omega)\sin(m\omega) \\ \sin(l\omega)\cos(m\omega) & \sin(l\omega)\sin(m\omega) \end{pmatrix} \quad (9.7)$$

with $\mathbf{R}$ the covariance matrix of the VAR process (Lütkepohl, 1993). rPDC is defined as

$$\lambda_{kj}(\omega) = \mathbf{X}_{kj}(\omega)'\mathbf{V}_{kj}^{-1}(\omega)\mathbf{X}_{kj}(\omega) , \quad (9.8)$$

where ()' means the matrix transposition. If $\lambda_{kj}(\omega) = 0$, a Granger-causal, linear influence from x_j to x_k taking all other processes into account can be positively excluded at the frequency ω. The α-significance level for $\lambda_{kj}(\omega) = 0$ and $\omega \neq 0$ is given by $\chi^2_{2,1-\alpha}/N$ (Schelter et al., 2009); rPDC is not normalized to [0, 1] as ordinary PDC (Equation 9.5).

If λ_{kj} exceeds the critical value at a certain frequency ω, process x_j is Granger-causal for x_k. Based on the knowledge of x_j, a prediction of x_k is possible to a certain extent. At a given frequency ω, rPDC $\lambda_{kj}(\omega)$ is a measure of the strength of an influence. Other than ordinary partial directed coherence, it is independent of outgoing influences from x_j to other processes than x_k. This way, two rPDC values at different frequencies are comparable.

9.3 TIME-DEPENDENT INFLUENCES

If the interaction between processes is not constant in time, rPDC analysis is expected to fail in estimating the true interaction structure. To demonstrate the effects of time-dependent influences, the two-dimensional VAR[2]-process

$$\vec{x}(t) = \sum_{r=1}^{2} \mathbf{a}_r \vec{x}(t-r) + \boldsymbol{\varepsilon}_x(t) \quad (9.9)$$

with

$$\mathbf{a}_1 = \begin{pmatrix} 1.3 & c \\ 0 & 1.7 \end{pmatrix}, \mathbf{a}_2 = \begin{pmatrix} -0.8 & 0 \\ 0 & -0.8 \end{pmatrix} \quad (9.10)$$

is investigated. The causal influence from process x_2 to x_1 is represented by the parameter c in our model, Equations 9.9 and 9.10. The time-dependent interaction is simulated by setting

$$c = \begin{cases} 0 & \text{if } t \leq k \\ 0.5 & \text{else} \end{cases}, \quad k = \left\{ 0, \frac{N}{6}, \dots, N \right\}, \quad (9.11)$$

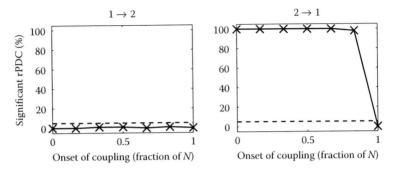

FIGURE 9.1 Significant rPDCs in % of 100 realizations for different onset times of coupling. rPDC was estimated using a model order of $p = 10$. Significance was evaluated at the oscillation frequencies of the potentially driving process, 0.12 and 0.05 Hz, respectively.

where N is the number of data points, that is, the influence from process x_2 to x_1 is present only in the second part of the simulation period. We generated 100 realizations of $N = 5000$ data points and used the Yule–Walker equations (Lütkepohl, 1993) with a model of order $p = 10$ for our estimation. Since the VAR[2]-processes used here correspond to driven damped oscillators with frequencies of approximately 0.12 and 0.05 Hz, respectively, rPDC was evaluated at these frequencies. In Figure 9.1, the percentage of significant rPDCs is shown depending on the different onset times of coupling.

Since the coefficients are estimated for the whole time series, rPDC could not detect this time-dependence of the coefficient c. The influence from process x_2 to x_1 evaluated at 0.05 Hz was detected in 100% of the simulated realizations, for all but one coupling scheme. For $k = N$, no coupling is present which is correctly revealed by rPDC. In summary, rPDC is able to detect a coupling even if present only for a short period of time. The drawback is that one cannot draw conclusions about the duration of the coupling.

Comparing the results for single realizations of the different coupling schemes (Equation 9.11), the estimated rPDC is higher for a longer duration of the coupling (Figure 9.2). This is consistent, since rPDC represents the mean coupling. A lower rPDC could thus correspond either to a shorter interaction or to a constant weaker coupling.

To distinguish between these two cases, a technique for time-resolved estimation is needed. Since coupling is piecewise constant, windowing the time series and estimating rPDC for each block is a possible solution in this scenario. Considering, for example, the case where the influence sets in at $k = N/2$, the transition is clearly visible in the time series, (Figure 9.3). Thus, the choice of blocks should account for the change in variance of the time series. For example, one can choose 20 nonoverlapping blocks; the results are shown in Figure 9.4. Choosing only two nonoverlapping blocks would suffice. This way, compared to 20 blocks, the frequency resolution could be optimized. Considering a different scenario, where the coupling increases linearly with time, that is, $c = \alpha t$, the time series do not indicate a specific choice for the block length, (Figure 9.5). In an application, one would have to face the challenge

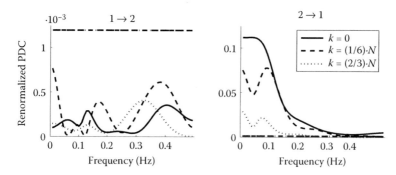

FIGURE 9.2 rPDCs of single realizations for different onset times of the coupling (Equation 9.11). rPDC was estimated using a model order of $p = 10$. The dashed dotted line corresponds to the 5%-significance level. Note the different scaling of the y-axis.

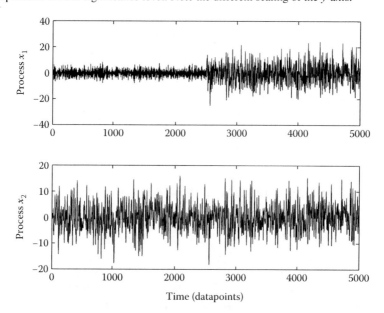

FIGURE 9.3 Time series of a single realization of VAR[2] with onset of coupling at $k = N/2$.

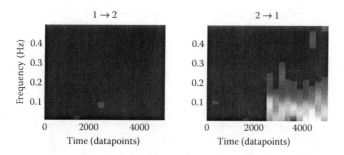

FIGURE 9.4 rPDCs of a single realization, onset of coupling at $k = N/2$. rPDC was estimated using a model order of $p = 10$.

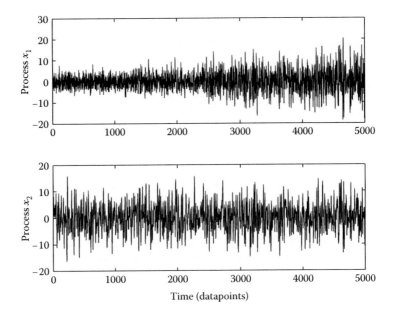

FIGURE 9.5 Time series of a single realization of VAR[2] with linearly increasing coupling.

of choosing an appropriate window length. The trade-off between high-frequency resolution, that is, long windows, and high time resolution, that is, short windows, has to be balanced.

Assuming another scenario, with an influence that is not piecewise constant as, for example,

$$c = \sin(25 \cdot t/N) \tag{9.12}$$

the windowing technique is not feasible. We simulated $N = 5000$ data points and estimated rPDC for 20 nonoverlapping segments with a model order of $p = 10$. The results are shown in Figure 9.6 for all frequencies and in Figure 9.7 for 0.12 Hz and 0.05, respectively. A change in the interaction structure is visible but the sinusoidal oscillation could not be reconstructed.

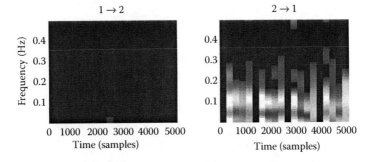

FIGURE 9.6 rPDCs for two AR[2] processes with oscillating coupling, Equation 9.12. rPDC was estimated using a model order of $p = 10$.

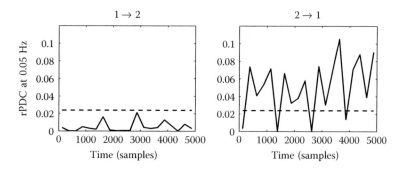

FIGURE 9.7 rPDCs for two AR[2] processes with oscillating coupling, Equation 9.12. rPDC was estimated using a model order of $p = 10$. The results are shown for oscillation frequencies of the potentially driving process, that is, 0.12 and 0.05 Hz, respectively.

In most applications, there is no prior knowledge about the time-dependence of the parameters. Therefore, a technique for time-resolved estimation is required.

9.4 KALMAN'S FILTER

To determine time-dependent rPDC, the parameter matrices $\mathbf{a}_1(t), \ldots, \mathbf{a}_p(t)$ of an n-dimensional time-dependent VAR process (Equation 9.1)

$$\vec{x}(t) = \sum_{r=1}^{p} \mathbf{a}_r(t)\vec{x}(t-r) + \boldsymbol{\varepsilon}_x(t), \qquad \boldsymbol{\varepsilon}_x(t) \sim \mathcal{N}(\vec{0}|\boldsymbol{\Sigma}), \qquad (9.13)$$

have to be estimated for each time step.

The linear state-space model is a powerful framework to estimate time-dependent parameters (Harvey, 1994; Kitagawa and Gersch, 1996). In the following, we present the concept of linear state-space modeling, first, in general; second, this state-space description of the VAR[p] process is generalized to time-dependent parameters $\mathbf{a}_1(t), \ldots, \mathbf{a}_p(t)$.

A stationary VAR process of order p and dimension n as introduced in Equation 9.1 can be rewritten as a process of first order by augmenting its dimension. Therefore, all past information needed to predict $\vec{x}(t)$ is collected in one single new $n_{\vec{u}}$-dimensional vector

$$\vec{u}(t-1) = \left(\vec{x}^T(t-1), \vec{x}^T(t-2), \ldots, \vec{x}^T(t-p)\right)^T, \qquad (9.14)$$

with dimension $n_{\vec{u}} = np$. The model equation

$$\underbrace{\begin{pmatrix} \vec{x}(t) \\ \vec{x}(t-1) \\ \vdots \\ \vec{x}(t-p+1) \end{pmatrix}}_{\vec{u}(t)} = \underbrace{\begin{pmatrix} \mathbf{a}_1 & \mathbf{a}_2 & \cdots & \mathbf{a}_p \\ \mathbf{I}_n & \mathbf{0}_n & \cdots & \mathbf{0}_n \\ \vdots & \ddots & \ddots & \vdots \\ \mathbf{0}_n & \cdots & \mathbf{I}_n & \mathbf{0}_n \end{pmatrix}}_{\mathbf{A}} \underbrace{\begin{pmatrix} \vec{x}(t-1) \\ \vec{x}(t-2) \\ \vdots \\ \vec{x}(t-p) \end{pmatrix}}_{\vec{u}(t-1)} + \underbrace{\begin{pmatrix} \boldsymbol{\varepsilon}_x(t) \\ \vec{0} \\ \vdots \\ \vec{0} \end{pmatrix}}_{\boldsymbol{\varepsilon}_u(t)}$$

$$(9.15)$$

with

$$\boldsymbol{\varepsilon}_u(t) \sim \mathcal{N} \left(\vec{0} \left| \underbrace{\left(\begin{array}{cccc} \boldsymbol{\Sigma} & \mathbf{0}_n & \cdots & \mathbf{0}_n \\ \mathbf{0}_n & \mathbf{0}_n & \cdots & \mathbf{0}_n \\ \vdots & \ddots & \ddots & \vdots \\ \mathbf{0}_n & \cdots & \mathbf{0}_n & \mathbf{0}_n \end{array} \right)}_{\mathcal{N}(\vec{0}|\mathbf{Q}_u)} \right. \right) \tag{9.16}$$

of the new vector $\vec{u}(t)$ is an equivalent representation of the VAR[p] process $\vec{x}(t)$. The matrices $\mathbf{I}_n$ and $\mathbf{0}_n$ denote the $n \times n$ identity and the $n \times n$ matrix of zeros. The new representation in Equation 9.15 of the VAR[p] process ensures a direct applicability of the linear state-space model (Shumway and Stoffer, 2000).

To include nonstationary dynamics, state-space modeling has to be extended to include time-dependent parameters $\mathbf{A}(t)$ with the matrix

$$\mathbf{A}(t) = \left(\begin{array}{cccc} \mathbf{a}_1(t) & \mathbf{a}_2(t) & \cdots & \mathbf{a}_p(t) \\ \mathbf{I}_n & \mathbf{0}_n & \cdots & \mathbf{0}_n \\ \vdots & \ddots & \ddots & \vdots \\ \mathbf{0}_n & \cdots & \mathbf{I}_n & \mathbf{0}_n \end{array} \right) \tag{9.17}$$

with $n^2 p$ time-varying entries of the coefficient matrices $\mathbf{a}_1(t), \ldots, \mathbf{a}_p(t)$ of the VAR[p] process.

To estimate time-dependent parameters $\mathbf{A}(t)$ using process $\vec{u}(t)$, the Kalman filter has been proposed (Wan and Nelson, 1997). Therefore, a state-space model

$$\mathbf{A}(t) = \mathbf{A}(t-1) + \boldsymbol{\varepsilon}_A(t), \tag{9.18}$$

$$\vec{u}(t) = \mathbf{C}\mathbf{A}(t) + \boldsymbol{\varepsilon}_u(t) \tag{9.19}$$

is used, which models the time-dependent parameters $\vec{A}(t)$ as a hidden process. Thereby, $\mathbf{C}$ denotes the observation matrix that results from the knowledge of $\vec{u}(t-1)$. The observation matrix $\mathbf{C}$ contains the process vector $\vec{u}(t-1)$ such that $\mathbf{C}\mathbf{A}(t) = \mathbf{A}(t)\vec{u}(t-1)$.

The Kalman filter comprises two steps. First, $\mathbf{A}(t|t-1)$ is predicted.

$$\mathbf{A}(t|t-1) = \mathbf{A}(t-1|t-1), \tag{9.20}$$

$$\mathbf{P}(t|t-1) = \mathbf{P}(t-1|t-1) + \mathbf{Q}, \tag{9.21}$$

$$\vec{u}(t|t-1) = \mathbf{C}(t)\mathbf{A}(t|t-1). \tag{9.22}$$

In the second step, the prediction is corrected according to the observation $\vec{u}(t)$

$$\mathbf{K}(t) = \mathbf{P}(t|t-1)\mathbf{C}^T(t) \left(\mathbf{C}(t)\mathbf{P}(t|t-1)\mathbf{C}^T(t) + \boldsymbol{\Sigma} \right)^{-1}, \tag{9.23}$$

$$\mathbf{P}(t|t) = (\mathbf{I}_{n^2 p} - \mathbf{K}(t)\mathbf{C}(t))\mathbf{P}(t|t-1), \tag{9.24}$$

$$\mathbf{A}(t|t) = \mathbf{A}(t|t-1) + \mathbf{K}(t)(\vec{u}(t) - \vec{u}(t|t-1)), \tag{9.25}$$

where $\mathbf{P}$ denotes variance of $\mathbf{A}$. The Kalman filter estimates the most likely trajectory, given the observations. This is not necessarily a typical trajectory of the process.

We emphasize that the apparently nonstationary equation

$$\mathbf{A}(t) = \mathbf{A}(t-1) + \boldsymbol{\varepsilon}_A(t) \tag{9.26}$$

for the parameter vector can be interpreted as a restriction of the difference $\mathbf{A}(t) - \mathbf{A}(t-1)$, which is the discrete counterpart of the first derivative, by independent Gaussian distributed noise $\boldsymbol{\varepsilon}_A(t) \sim \mathcal{N}(\vec{0}|\mathbf{Q})$ with a diagonal covariance matrix $\mathbf{Q}$ (Kitagawa and Gersch, 1996). In order to differentiate between the process dynamics and the changes in the parameter $a_i(t)$, the latter is allowed to change only up to a certain amount in each time step. The variance Q_{ii} determines this amount and thus the smoothness of the entire parameter curve $a_i(t)$. The Kalman filter computes the optimal estimates $\vec{a}(t|t)$, given the process state vectors $\vec{u}(t-1)$. If the parameters $\mathbf{Q}$ and $\boldsymbol{\Sigma}$ are unknown, they can be optimized using the expectation–maximization algorithm (Honerkamp, 2012; Sommerlade et al., 2012)

9.5   KALMAN'S FILTER AND RPDC

Since rPDC is defined by the p coefficient matrices of the VAR[p] process underlying the time series and their covariances, the determination of time-resolved rPDC corresponds to the estimation of a time-variant VAR[p] process as proposed in the last section (Killmann et al., 2012). Once the p VAR coefficient matrices and their covariances are estimated $\{\mathbf{a}_1(t), \ldots, \mathbf{a}_p(t)\}$, rPDC can be calculated for every time step t using Equation 9.8.

As an example, we use the two-dimensional VAR[2]-process, Equation 9.9, with the coupling parameter c given in Equation 9.11. The results for rPDC in combination with the Kalman filter are shown in Figure 9.8. For the estimation, a model order of $p = 10$ was used. The onset of the interaction was correctly revealed.

rPDC estimated for segments of the data is not able to reveal a sinusoidal oscillation of the simulated coupling (Section 9.3). In combination with the Kalman filter, this time dependence can be revealed. The results are shown in Figure 9.9 for all frequencies and in Figure 9.10 for the oscillation frequencies of the potentially driving processes, 0.12 and 0.05 Hz, respectively. The oscillating coupling, Equation 9.12, is reconstructed correctly by this method.

To illustrate the multivariate capability of the method presented here, we simulated a four-dimensional VAR[2]-process with $N = 5000$ data points and parameters

$$\mathbf{a}_1 = \begin{pmatrix} 1.3 & c_{12} & 0 & c_{14} \\ c_{21} & 1.6 & 0 & 0 \\ c_{31} & 0 & 1.5 & c_{34} \\ 0 & 0 & 0 & 1.7 \end{pmatrix} \tag{9.27}$$

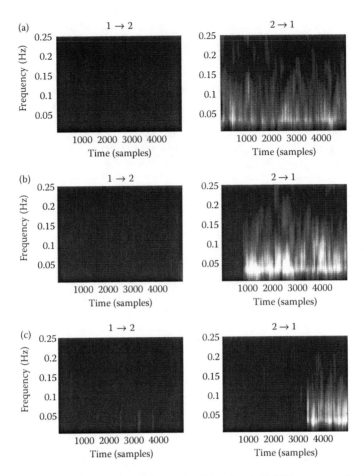

FIGURE 9.8 Time-dependent rPDC for a two-dimensional VAR[2] process with different onset times of coupling. (a) Interaction sets in at the beginning of the simulation. (b) Onset time of $k = 1/6 \cdot N$. (c) Onset time of $k = 2/3 \cdot N$. On the left, rPDC for the direction from x_1 to x_2 is shown. In the simulation, only the coupling from x_2 to x_1 is present, for which the results are shown on the right. Color coding (gray scale) indicates the intensity of coherence.

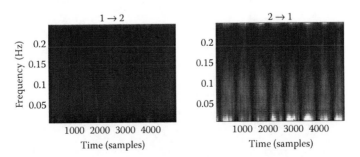

FIGURE 9.9 Time-dependent rPDC for the two-dimensional VAR[2]-process with nonstationary coupling. On the left, rPDC for the direction from x_1 to x_2 is shown. In the simulation, the coupling from x_2 to x_1 is oscillating according to Equation 9.12, for which the results are shown on the right. Color coding (gray scale) indicates the intensity of coherence.

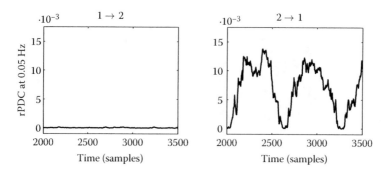

FIGURE 9.10 Time-dependent rPDC for a segment of the two-dimensional VAR[2]-process with nonstationary coupling. rPDC for the direction from x_1 to x_2 at 0.12 Hz and from x_2 to x_1 at 0.05 Hz is shown. In the simulation, the coupling is oscillating according to Equation 9.12.

$$\mathbf{a}_2 = \begin{pmatrix} -0.8 & 0 & 0 & 0 \\ 0 & -0.8 & 0 & 0 \\ 0 & 0 & -0.8 & 0 \\ 0 & 0 & 0 & -0.8 \end{pmatrix} \tag{9.28}$$

and

$$c_{31} = 0.5, \tag{9.29}$$

$$c_{12} = \begin{cases} 0 & \text{if } t \leq 3333, \\ 0.7 & \text{else} \end{cases} \tag{9.30}$$

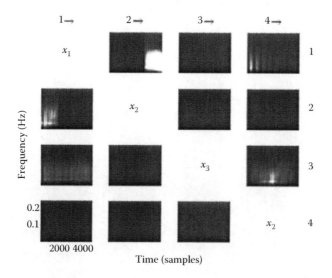

FIGURE 9.11 Results of rPDC analysis in combination with the Kalman filter for the four-dimensional VAR[2]-process, Equations 9.27 through 9.33. Color coding (gray scale) indicates the intensity of coherence.

$$c_{21} = \begin{cases} 0.7 & \text{if } t \leq 1666, \\ 0 & \text{else} \end{cases} \tag{9.31}$$

$$c_{34} = 0.8 \left(1 - \frac{|t - 2500|}{2500} \right), \tag{9.32}$$

$$c_{14} = e^{-t/2500} \sin(0.005t). \tag{9.33}$$

This corresponds to the following interaction structure: The interaction from x_1 to x_3 is constant for the whole simulation; the interaction from x_1 to x_2 is present only in the first third of the simulation while the reverse direction (x_2 to x_1) is present in the last third. The interaction from x_4 to x_3 starts with zero, is linearly increased until the middle of the simulation, and then linearly decreases to zero until the end of the simulation. The interaction from x_4 to x_1 is modulated by a damped oscillation decaying to zero at the end of the simulation. All other interactions are zero. The results using a model order of $p = 10$ for this simulation are shown in Figure 9.11. All simulated interactions are revealed correctly. Thus, the method presented here is capable of inferring the time-dependent interaction structure in multivariate systems.

9.6 APPLICATION

We next applied the proposed approach to a biological data set, based on electroencephalographic (EEG) data acquired from a mouse during a transition from slow-wave (NREM) to rapid eye movement (REM) sleep. EEG recordings from three locations were obtained under freely moving conditions, that is, in a prefrontal position (above the right prefrontal cortex) and in two parietal locations above the left and right hippocampus (for details of the recording technique, see Jyoti et al., 2010) using a wireless device (sampling rate: 199 Hz). Figure 9.12 depicts raw data (diagonal graphs) as well as plots for rPDC; the transition takes place after 16.7 s as indicated by a vertical line. Coherence data suggest stronger interactions between all brain regions during REM compared to NREM sleep; during the latter stage, only intermittent and variable interactions were detected. Coincident with the onset of REM, coherence occurred particularly at frequencies between 10 and 20 Hz, thus comprising the alpha and beta range. In the prominent alpha range (i.e., 9–14 Hz), an influence from the left hippocampus (lHC) to the prefrontal cortex (PFx) occurred earliest in this example. The opposite direction (PFx to lHC) followed and strengthened toward the end of the recording. The onset of the influence from the right hippocampus (rHC) to the left hippocampus (lHC) was strongest and mirrored in the lHC to PFx connection. Interestingly, though the interactions from lHC to rHC as well as from rHC to PFx were weaker, they increased toward the end of the recording.

Capabilities to reliably detect and predict the behavioral and vigilance stages are currently limited. Thus, time-resolved rPDC can now provide advanced measures for connectivity between brain areas and identify EEG-based markers for the underlying neuronal processes and networks. The procedure presented here may enable a better understanding of directed communication within networks that determine

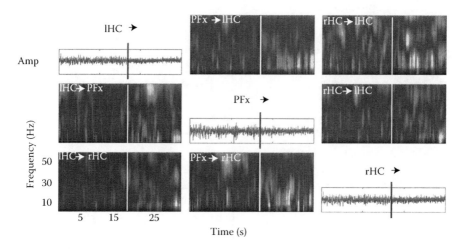

FIGURE 9.12 **(See color insert.)** rPDC analysis applied to murine EEG data recorded bilaterally above the hippocampus (left: lHC, right: rHC), and prefrontal cortex (Pfx). Raw data are depicted on the diagonal (amplitude (Amp) in arbitrary units) over time in seconds. The lines indicate the time point of transition between NREM to REM sleep. For the rPDC spectra, direction of information flow is from column to row. Color coding (gray scale) indicates the intensity of coherence.

specific behaviors and cognitive processes. This will be rigorously tested in future investigations to provide novel analytical tools for a range of applications.

9.7   DISCUSSION AND CONCLUSION

Detection of relationships and causal influences is of special interest in multivariate time-series analysis. In many applications, this task is hampered by time-dependent parameters of processes and especially relationships between processes. We presented a combination of renormalized partial directed coherence and the Kalman filter, which is feasible in detecting time-variant causal influences in multivariate processes.

Using just the Kalman filter often leads to hardly interpretable results, if too many coefficients are unequal to zero. This difficulty is overcome by renormalized partial directed coherence. Moreover, renormalized partial directed coherences are estimated in the frequency domain. This allows detection of causal influences at characteristic frequencies, which is advantageous for oscillatory processes in applications. Furthermore, a time-dependency in the parameter estimation is included in the analysis. This is superior to cutting time series into epochs and estimating parameters for each epoch, which by construction yields piecewise constant parameter estimates. Additionally, the cutting points force the parameter estimation technique to assume stationarity for each epoch. This might lead to results depending on the choice of cut points. Additionally, the length of the segments has to be selected. Thereby, the trade-off between a high-frequency resolution, that is, long segments, and a high temporal resolution, that is, short segments, has to be faced.

The combination and extension of renormalized partial directed coherence and the Kalman filter allows the detection of time-dependent and possibly rapidly changing causal influences in multivariate linear processes. Apart from nonstationarity, the challenge of observational noise has to be faced in applications. State-space modeling presents a good framework to account for observational noise. Therefore, we suggest an additional observation equation in the state-space model and estimate the coefficients based on this model.

Renormalized partial directed coherence in combination with the Kalman filter and EM algorithm enables a time-resolved estimation of Granger-causal interaction structures from measured data. Thus, it promises deeper insights into mechanisms underlying network phenomena.

ACKNOWLEDGMENTS

This work was supported by the German Science Foundation (Ti315/4-2), the German Federal Ministry of Education and Research (BMBF grant 01GQ0420), and the Excellence Initiative of the German Federal and State Governments. B.S. and L.S. are indebted to the Landesstiftung Baden-Württemberg for the financial support of this research project by the Eliteprogramme for postdocs. This work was partially funded by the UK Biotechnology and Biological Sciences Research Council under the Systems Approaches to Biological Research (SABR) Initiative (BB/FO0513X/1). B.S., B.P., and M.T. acknowledge support from the Kosterlitz Centre Pump Priming Fund.

REFERENCES

Arnold, M., W. H. R. Miltner, H. Witte et al. 1998. Adaptive AR modeling of nonstationary time series by means of Kalman filtering. *IEEE Trans. Biomed. Eng.* 45: 553–562.

Baccalá, L. A. and K. Sameshima. 1998. Directed coherence: A tool for exploring functional interactions among brain structures. In M. Nicolelis (Ed.), *Methods for Neural Ensemble Recordings*, 179–192. Boca Raton: CRC.

Baccalá, L. A. and K. Sameshima. 2001. Partial directed coherence: A new concept in neural structure determination. *Biol. Cybern.* 84: 463–474.

Baccalá, L. A., K. Sameshima, G. Ballester et al. 1998. Studying the interaction between brain structures via directed coherence and Granger causality. *Appl. Signal Process.* 5: 40–48.

Brillinger, D. R. 1981. *Time Series: Data Analysis and Theory.* San Francisco: Holden-Day.

Brockwell, P. J. and R. A. Davis. 1998. *Time Series: Theory and Methods.* New York: Springer.

Dahlhaus, R. 2000. Graphical interaction models for multivariate time series. *Metrika 51*: 157–172.

Dahlhaus, R. and M. Eichler. 2003. Causality and graphical models for time series. In P. Green, N. Hjort, and S. Richardson (Eds.), *Highly Structured Stochastic Systems*, 115–137. Oxford: Oxford University Press.

Eichler, M. 2000. Markov properties for graphical time series models. *Preprint.* University of Heidelberg.

Geweke, J. 1982. Measurement of linear dependence and feedback between multiple time series. *J. Am. Stat. Assoc. 77*: 304–313.

Geweke, J. 1984. Measures of conditional linear dependence and feedback between time series. *J. Am. Stat. Assoc. 79*: 907–915.

Granger, J. 1969. Investigating causal relations by econometric models and cross-spectral methods. *Econometrica 37*: 424–438.

Grenier, Y. 1983. Time-dependent ARMA modeling of nonstationary signals. *IEEE Trans. Acoust. Speech 31*: 899–911.

Halliday, D. M. and J. R. Rosenberg. 2000. On the application and estimation and interpretation of coherence and pooled coherence. *J. Neurosci. Methods 100*: 173–174.

Hamilton, J. D. 1994. *Time Series Analysis.* Princeton: Princeton University Press.

Harvey, A. C. 1994. *Forecasting Structural Time Series Models and the Kalman filter.* Cambridge: Cambridge University Press.

Honerkamp, J. 1993. *Stochastic Dynamical Systems.* New York: VCH.

Honerkamp, J. 2012. *Statistical Physics: An Advanced Approach with Applications.* Heidelberg: Springer.

Jyoti, A., A. Plano, G. Riedel et al. 2010. EEG, activity, and sleep architecture in a transgenic $A\beta PP_{swe}/PSEN1_{A246E}$ Alzheimer's disease mouse. *J. Alzheimers Dis. 22*: 873–887.

Kamiński, M. J. and K. J. Blinowska 1991. A new method of the description of the information flow in the brain structures. *Biol. Cybern. 65*: 203–210.

Kamiński, M. J., K. J. Blinowska, and W. Szelenberger. 1997. Topographic analysis of coherence and propagation of EEG activity during sleep and wakefulness. *Electroenceph. clin. Neurophysiol. 102*: 216–227.

Killmann, M., L. Sommerlade, W. Mader et al. 2012. Inference of time-dependent causal influences in networks. *Biomed. Eng. 57*: 387–390.

Kitagawa, G. and W. Gersch. 1996. *Smoothness Priors Analysis of Time Series.* New York: Springer.

Korzeniewska, A., S. Kasicki, M. J. Kamiński et al. 1997. Information flow between hippocampus and related structures during various types of rat's behavior. *J. Neurosci. Methods 73*: 49–60.

Lütkepohl, H. 1993. *Introduction to Multiple Time Series Analysis.* New York: Springer.

Nicolelis, M. A. L. and E. E. Fanselow. 2002. Thalamocortical optimization of tactile processing according to behavioral state. *Nat. Neurosci. 5*: 517–523.

Pompe, B., P. Blidh, D. Hoyer et al. 1998. Using mutual information to measure coupling in the cardiorespiratory system. *IEEE Eng. Med. Biol. Mag. 17*: 32–39.

Rosenblum, M. G., L. Cimponeriu, A. Bezerianos et al. 2002. Identification of coupling direction: Application to cardiorespiratory interaction. *Phys. Rev. E 65*: 041909.

Rosenblum, M. G. and A. S. Pikovsky. 2001. Detecting direction of coupling in interacting oscillators. *Phys. Rev. E 64*, 045202(R).

Sameshima, K. and L. A. Baccalá. 1999. Using partial directed coherence to describe neuronal ensemble interactions. *J. Neurosci. Methods 94*: 93–103.

Schack, B., G. Grießbach, M. Arnold et al. 1995. Dynamic cross spectral analysis of biological signals by means of bivariate ARMA processes with time-dependent coefficients. *Med. Biol. Eng. Comput. 33*: 605–610.

Schelter, B., J. Timmer, and M. Eichler. 2009. Assessing the strength of directed influences among neural signals using renormalized partial directed coherence. *J. Neurosci. Methods 179*: 121–130.

Schelter, B., M. Winterhalder, M. Eichler et al. 2005. Testing for directed influences among neural signals using partial directed coherence. *J. Neurosci. Methods 152*: 210–219.

Shumway, R. H. and D. S. Stoffer. 2000. *Time Series Analysis and its Application.* New York: Springer.

Sommerlade, L., M. Thiel, B. Platt et al. 2012. Inference of Granger causal time-dependent influences in noisy multivariate time series. *J. Neurosci. Methods 203*: 173–185.

Wan, E. A. and A. T. Nelson. 1997. Neural dual extended Kalman filtering: Applications in speech enhancement and monaural blind signal separation. *IEEE Proceedings in Neural Networks for Signal Processing: VII. Proceedings of the 1997 IEEE Workshop*, 466 – 475.

Wan, E. A. and R. van der Merwe. 2000. The unscented Kalman filter for nonlinear estimation. *IEEE Communications and Control Symposium for Adaptive Systems for Signal Processing*, 153–158.

Section III

Applications

10 Connectivity Analysis Based on Multielectrode EEG Inversion Methods with and without fMRI A Priori Information

Laura Astolfi and Fabio Babiloni

CONTENTS

10.1 INTRODUCTION

Human neocortical processes involve temporal and spatial scales spanning several orders of magnitude, from the rapidly shifting somatosensory processes, characterized by a temporal scale of milliseconds and spatial scales of a few square millimeters, to the memory processes, involving time periods of seconds and spatial scales of square centimeters. Information about brain activity can be obtained by measuring different physical variables arising from the brain processes, such as the increase in consumption of oxygen by the neural tissues or the variation of the electric potential on the scalp surface. All these variables are connected directly or indirectly to the neural ongoing processes, and each variable has its own spatial and temporal resolution. The different neuroimaging techniques are then confined to the spatio-temporal resolution offered by the monitored variables. For instance, it is known from physiology that the temporal resolution of the hemodynamic deoxyhemoglobin increase/decrease lies in the range of 1–2 s, while its spatial resolution is generally observable with the current imaging techniques over the scale of a few millimeters.

Electroencephalography (EEG) and magnetoencephalography (MEG) are two techniques that present high temporal resolution, on the millisecond scale, adequate to follow brain activity related to cognitive processes. Unfortunately, both techniques have a relatively modest spatial resolution, beyond the centimeter, being fundamentally limited by the inter-sensor distances and by the fundamental laws of electromagnetism (Nunez, 1981).

However, a body of techniques were developed in the last 20 years to increase the spatial resolution obtainable from neuroelectrical recordings. Such techniques include the use of a large number of scalp sensors, realistic models of the head derived from structural magnetic resonance images (MRIs), and advanced processing methods related to the solution of the linear inverse problem. These methods allow the estimation of cortical current density from sensor measurements (de Peralta Menendez and Gonzalez Andino, 1999; Babiloni et al., 2000).

On the other hand, the use of a priori information from other neuroimaging techniques such as functional magnetic resonance imaging (fMRI) with high spatial resolution could improve the localization of sources from EEG/MEG data. The rationale of the multimodal approach based on the fMRI, MEG, and EEG data to locate brain activity is that neural activity generating EEG potentials or MEG fields increases glucose and oxygen demands (Magistretti et al., 1999). This results in an increase in the local hemodynamic response that can be measured by fMRI (Grinvald et al., 1986; Puce et al., 1997). On the whole, such a correlation between electrical and hemodynamic concomitants provides the basis for the spatial correspondence between fMRI responses and EEG/MEG source activity. Scientific evidence suggests that the estimation of the cortical activity performed with the use of neuroelectromagnetic recordings improves with the use of hemodynamic information recording during the same task (Nunez, 1981; Gevins, 1989; Urbano et al., 1998, 2001; Dale et al., 2000; He and Lian, 2002; Babiloni et al., 2005; He et al., 2006; Im et al., 2006). This increase in the quality of the estimation of cortical activity was also assessed during simulation studies that clearly demonstrated the advantage of the inclusion of the functional magnetic resonance imaging (fMRI) priors into the cortical estimation process (Dale et al., 2000; Babiloni et al., 2003, 2004; Im et al., 2006).

The improvement of the spatio-temporal resolution of the cortical estimates obtained by the inclusion of the information given by fMRI to the linear inverse estimation has an expected impact on the estimation of functional connectivity. Spectral techniques to determine the directional influences between any given pair of channels in a multivariate data set like those based on Granger causality, partial directed coherence (PDC), directed transfer function (DTF), and their different variations and normalizations (Kamiński and Blinowska, 1991; Baccalá and Sameshima, 2001; Kamiński et al., 2001) were applied to cortical reconstruction data sets obtained from high-resolution EEG recordings since 2005 (Astolfi et al., 2005, 2006b) and to cortical reconstruction obtained with the inclusion of fMRI information in the solution of the linear inverse estimation of the cortical activity (Babiloni et al., 2005; Astolfi et al., 2006a, 2007).

Since the organization of brain circuits is thought to be based on the interaction between different and differently specialized cortical sites, the precise localization of such cortical areas is crucial for the estimation of relevant cortical circuits. Moreover, the linear inverse procedure aims at removing the correlation induced to the scalp EEG data by the effect of volume conduction. In addition, it makes the description of the cortical circuits more straightforward by giving a clear interpretation of which cortical sites are involved. Finally, the reduction from the electrode domain to the cortical domain, where regions of interest may be drawn and selected according to the specific experimental task of interest, may help reduce the number of time series to be modeled by the unique MVAR model, which is crucial in the multivariate approach.

Simulation studies were performed in which factors affecting the EEG recordings (such as the signal-to-noise ratio, the length of the recordings, and the existence of direct and indirect links) were systematically changed to study their effect on different indexes of the performances (the general relative error, the error on a specific model link, the number of false positives and false negatives). Such performances were evaluated by an analysis of variance (ANOVA) revealing the range of conditions showing acceptable performances by the estimators (Astolfi et al., 2005, 2006a,b). The results showed that, under conditions usually met in this kind of experiments, the results provided by the application of these methodologies are reliable, with an error below 5%.

Finally, we also present both the mathematical principle and the practical applications of the multimodal integration of high-resolution EEG and fMRI for the localization of sources responsible for intentional movements.

In this chapter, we will review the procedure for the estimation of cortical connectivity from linear inverse estimates of the cortical activity obtained from the scalp EEG data with and without the additional spatial information given by fMRI measurements.

Comparisons are made between the connectivity patterns returned, for the same data set, by the different spectral multivariate estimators (DTF, PDC, dDTF), and the results obtained by EEG data and by EEG-fMRI integration.

10.2 RECONSTRUCTION OF THE CORTICAL ACTIVITY

To estimate the cortical activity from neuroelectromagnetic measurements by using the distributed source approach, a lead field matrix A has to be computed for each EEG recording performed. Such a matrix depends critically on the sensors' positions as well as on the actual geometry of the subject's head. The geometrical information about the structures of the subject's head can be obtained from structural magnetic resonance images. By analyzing such images, we can build tessellated surfaces of the cortical surface, of the inner and outer skull surfaces, and of the scalp surface. Conductivity of the skull compartment is usually set to 15 ms/m; conductivity of both the scalp and the brain compartments is around 15 times higher than that of the

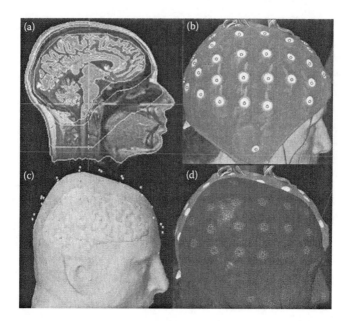

FIGURE 10.1   Generation of the Lead Field matrix for the estimation of cortical current density from high-resolution EEG recordings. (a) MRI images from a healthy subject, (c) generation of the head models and superimposition (b), (d) with an electrodes cap.

skull. Such values were indicated by *in vivo* anatomical studies. The cortical sources can be modeled using a distributed model with a realistic cortical shape (Babiloni et al., 2000). A tessellated surface representing the cortical mantle (about 5000–9000 triangles on average) is extracted through an iterative procedure, which fits it halfway between the white/gray and the gray/CFS interfaces. The source model is thus composed by 4000–8000 current dipoles, positioned at the vertices of the cortical tessellation, with direction normal to the local pseudo-tangent plane. Different cortical regions of interest (ROIs) can be drawn on the computer-based cortical reconstruction of the individual head models, based on the anatomical landmarks and on the Brodmann areas. The steps of the construction of the Lead Field Matrix are illustrated by Figure 10.1.

10.2.1   ESTIMATION OF THE CORTICAL SOURCE CURRENT DENSITY

The solution of the following linear system:

$$\mathbf{Ax} = \mathbf{b} + \mathbf{n} \tag{10.1}$$

provides an estimate of the dipole source configuration $\mathbf{x}$ that generated the measured EEG potential distribution $\mathbf{b}$. The system also includes the measurement noise $\mathbf{n}$, which is supposedly normally distributed.

In Equation 10.1, $\mathbf{A}$ is the lead field matrix, in which each j-th column describes the potential distribution generated on the scalp electrodes by the j-th unitary dipole. The current density solution vector $\boldsymbol{\xi}$ is obtained as (de Peralta Menendez and Gonzalez

Andino, 1999)

$$\xi = \arg\min_{x} (\| \mathbf{A}x - \mathbf{b} \|_{\mathbf{M}}^2 + \lambda^2 \| \mathbf{x} \|_{\mathbf{N}}^2), \tag{10.2}$$

where $\mathbf{M}, \mathbf{N}$ are the matrices associated with the metrics of the data and of the source space, respectively, λ is the regularization parameter, and $\|\mathbf{x}\|_M$ represents the M norm of the vector $\mathbf{x}$. The solution of Equation 10.2 is given by the inverse operator $\mathbf{G}$ as follows:

$$\xi = \mathbf{G}\mathbf{b}, \mathbf{G} = \mathbf{N}^{-1}\mathbf{A}'(\mathbf{A}\mathbf{N}^{-1}\mathbf{A}' + \lambda\mathbf{M}^{-1})^{-1}. \tag{10.3}$$

An optimal regularization of this linear system was obtained by the L-curve approach (Hansen, 1992a,b). As a metric in the data space, we used the identity matrix, while as a norm in the source space we used the following metric, which takes the hemodynamic information offered by the recorded fMRI data into account

$$(\mathbf{N}^{-1})_{ii} = g(\alpha)^2 \| \mathbf{A}_{\cdot i} \|^{-2}, \tag{10.4}$$

where $(\mathbf{N}^{-1})_{ii}$ is the i-th element of the inverse of the diagonal matrix $\mathbf{N}$ and all the other matrix elements $\mathbf{N}_{ij}$ are set to 0. The L_2 norm of the i-th column of the lead field matrix $\mathbf{A}$ is denoted by $\|\mathbf{A}_{\cdot i}\|$. The $g(\alpha_i)$ is a function of the statistically significant percentage increase of the fMRI signal during the task, compared to the rest state. Such a function has values greater than 1 for positive α_i, while it takes values lower than 1 for negative α_i. In particular, such a function is defined as

$$g(\alpha) = \max \left\{ 1 + (K-1)\frac{\alpha_i}{\max\limits_{i}\{\alpha_i\}}, \ ^1\!/K \right\}, \quad K \geq 1, \ \forall \alpha, \tag{10.5}$$

where the value of the parameter K tunes the strength of the inclusion of the fMRI constraints in the source space, and the function $\max(a, b)$ takes the maximum of the two arguments for each α. Here, we used the value of $K = 10$, which resulted from a previous simulation study, as a value returning optimal estimation of source current density with fMRI priors for a large range of SNR values of the gathered EEG signals (Babiloni et al., 2003).

10.3 CONNECTIVITY ESTIMATION

In this chapter, we will briefly discuss some methodological issues related to the application of DTF, PDC, and dDTF (whose full mathematical description has been provided previously in this book) to the cortical signals obtained from the application of the inverse procedures described above.

10.3.1 STATISTICAL EVALUATION OF CONNECTIVITY MEASUREMENTS

All the methods for the estimation of the cortical connectivity return a series of values for the causal links between the ROIs analyzed. The crucial issue then is whether these values are significantly different from chance or not.

A possible approach to assess the statistical significance of the MVAR-derived connectivity methods (DTF, PDC, dDTF) consists in the use of a surrogate data technique (Theiler et al., 1992), in which an empirical distribution of the null case is generated by evaluating connectivity from several realizations of surrogate data sets, in which deterministic interdependency between variables was removed. In order to ensure that all the features of each data set are as similar as possible to the original data set, with the exception of channel coupling, the very same data are used, and any time-locked coupling between channels is disrupted by shuffling the phases of the original multivariate signal. In this way, the temporal relationship among channels is destroyed while the spectral properties of the signals are preserved, thus returning a different threshold for different frequencies. The best approach is to shuffle the phases of each time series separately, while keeping all the other series as in the original data, to build the specific threshold for all the links related to that series, and then to repeat the procedure for all the time series in the data set.

This procedure is quite time-consuming. Recently, a validation approach based on the asymptotic distributions of PDC was proposed (Schelter et al., 2006; Takahashi et al., 2007). This new method is based on the assumption that the PDC estimator tends to a χ^2 distribution in the null case (lack of transmission) and to the normal distribution for the non-null hypothesis. Based on this assumption, statistical tests are possible to assess the significance of the connectivity patterns obtained.

10.4   APPLICATION TO REAL DATA

In this section, application examples of the methods described in the previous paragraphs on a series set of data from simple motor and cognitive tasks are presented.

10.4.1   COMPARISON BETWEEN DIFFERENT FUNCTIONAL CONNECTIVITY ESTIMATORS

In this section, we will show an example of the results obtained by the application of different functional connectivity estimators (DTF, PDC, and dDTF) on the same data set.

The application was performed on data recorded from a population of healthy volunteers during a cognitive task, the Stroop color-word interference test (Stroop, 1935). In this task, subjects are put in front of a screen where some words are presented to them. They have to indicate, in the shortest time they can, the color in which each word is written. According to the meaning of the word, there are three possible conditions: (1) when the word meaning is a color (e.g., red) and is presented in the same color (e.g., in red), we are in a Congruent condition; (2) if the word means a color (red) but is presented in a different color (e.g., in blue), we are in the Incongruent condition; and (3) if the word meaning is not related to colors, we are in the Neutral condition. Blocks of congruent or incongruent words are alternated with blocks of neutral words. A trial begins with the presentation of a word for 1500 ms, followed by a fixation cross for an average of 500 ms (randomly presented). Each trial consists of one word presented in one of four colors (red, yellow, green, blue), with each color occurring equally often with each word type (congruent, neutral, incongruent).

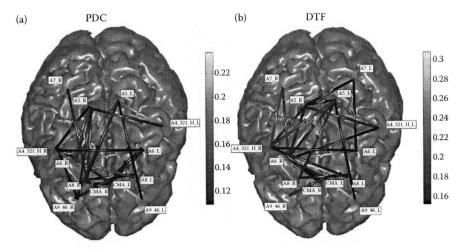

FIGURE 10.2 (**See color insert.**) Cortical connectivity patterns obtained by PDC (a) and DTF (b) for the period preceding the subject's response during congruent trials in the beta (12–29 Hz) frequency band. The patterns are shown on the realistic head model and cortical envelope of the subject, obtained from sequential MRIs. The brain is seen from above, left hemisphere represented on the right side. The colors and sizes of the arrows code the level of strengths of the cortical connectivity observed among ROIs. The lighter and the bigger the arrows, the stronger the connections. Only the cortical connections statistically significant at $p < 0.01$ are reported.

Subjects are asked to press one of four buttons corresponding to the color of the ink of the presented word. Data from 0 to 450 ms poststimulus onset were analyzed. Behavioral data (Reaction Times) were also collected.

Figure 10.2 shows the cortical connectivity patterns obtained for the period preceding the subject's response during the congruent trials in the beta (12–29 Hz) frequency band, for a representative subject. The patterns are shown on the realistic head model and cortical envelope of the subject, obtained from MRI scans. Two connectivity patterns are depicted, estimated in the beta frequency band for the same subject with the PDC (a), and the DTF (b). Only the statistically significant cortical connections are reported. It is possible to observe that in this case the cortical estimated patterns were preserved with slight differences between DTF and PDC, whereas the main structure of the circuit was preserved. In particular, functional connections between parietofrontal areas were present in all the estimations performed by the DTF, PDC, and dDTF methods. Moreover, connections involving the cingulate cortex are also clearly visible, as well as those involving the prefrontal areas, mainly in the right hemisphere. Functional connections in the prefrontal and premotor areas tended to be greater in the right hemisphere, whereas the functional activity in the parietal cortices was generally bilateral. It is still possible to observe the substantial agreement of the patterns depicted by the different methods for the estimation of the cortical patterns with the different functional connectivity estimators.

The connectivity patterns in the different frequency bands can also be summarized by using indices representing the total flow from and toward the selected cortical area. The total inflow in a particular cortical region i is defined as the sum of all the statistically significant connections from all the other cortical regions toward i. The total inflow for each ROI is represented by a sphere centered on the cortical region, whose radius is linearly related to the magnitude of all the incoming statistically significant links from the other regions. This information depicts each ROI as the target of functional connections from the other ROIs. The same conventions can be used to represent the total outflow from a cortical region, generated by the sum of all the

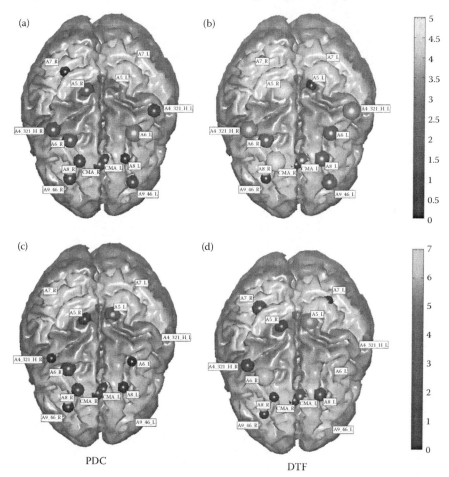

FIGURE 10.3 **(See color insert.)** The inflow (a) and (b) and the outflow (c) and (d) patterns obtained for the beta frequency band from each ROI during the congruent trials. The brain is seen from above, left hemisphere represented on the right side. Different ROIs are depicted in colors (selected regions: left and right primary motor areas (BA4); posterior supplementary motor areas (pSMA: BA6); right and left parietal areas including BA7 and 5; Bas 8 and 9/46; left and right cingulate motor areas (CMA). (a) and (b): in red hues, the behavior of an ROI as the target of information flow from other ROIs. The larger the sphere, the higher the value of inflow or outflow for any given ROI. (c) and (d): in blue hues, the outflow of information from a single ROI toward all the others.

statistically significant links spreading from the ROI toward the others. Figure 10.3 shows the inflow (first row) and outflow (second row) patterns computed, for the same subject of Figure 10.2, in the beta frequency band, by the PDC and DTF methods. The ROIs that are very active as source or sink (i.e., the source/target of the information flow to/from other ROIs) show results that are generally stable across the different estimators. In fact, a major involvement of the right premotor and prefrontal regions is observed across methods.

10.4.2   COMPARISON BETWEEN FUNCTIONAL CONNECTIVITY ESTIMATES PERFORMED WITH AND WITHOUT THE MULTIMODAL INTEGRATION OF NEUROELECTRIC AND HEMODYNAMIC INFORMATION

In this section, we present a comparison between the results of the estimation of cortical connectivity performed by DTF computed on the linear inverse estimation based on (i) EEG data only and (ii) EEG with fMRI *priors*. We will show two examples, one related to a finger-tapping paradigm and the other from the same Stroop's task data discussed in the previous section.

10.4.2.1   Finger-Tapping Data Acquisition

The experiment is described in detail in Babiloni et al. (2005) and Astolfi et al. (2005). Subjects were seated comfortably in an armchair with both arms relaxed and resting on pillows and were requested to perform fast repetitive right finger movements that were cued by visual stimuli. Ten to fifteen blocks of 2 Hz thumb oppositions for right hands were recorded with each 30-s blocks of finger movement and rest. During the movement, subjects were instructed to avoid eye blinks, swallowing, or any other motor activity other than the required finger movement. Event-related potential (ERP) data were recorded with 96 electrodes; data were recorded with a left ear reference and submitted to the artifact removal processing. Six hundred ERP trials of 600 ms of duration were acquired. The A/D sampling rate was 250 Hz. The surface electromyographic (EMG) activity of the right arm muscles was also collected. The onset of EMG response served as zero time. All data were visually inspected, and trials containing artifacts were rejected. We used a semiautomatic supervised threshold criteria for the rejection of trials contaminated by ocular and EMG artifacts. After the EEG recording, the electrode positions were digitized using a three-dimensional localization device with respect to the anatomic landmarks of the head (nasion and two preauricular points). The MRIs of each subject's head were also acquired. These images were used for the construction of the individual realistic geometry head model. The spectral values of the estimated cortical activity in the theta (4–7 Hz), alpha (8–12 Hz), beta (13–30 Hz), and gamma (30–45 Hz) frequency bands were also computed in each defined ROI. The analysis period for the potentials time locked to the movement execution was set from 300 ms before to 300 ms after the EMG trigger (0 time); the ERP time course was divided into two phases relative to the EMG onset; the first, labeled as the PRE period, marked the 300 ms before the EMG onset and was intended as a generic preparation period; the second, labeled as the POST period, lasted up to 300 ms after the EMG onset and was intended to signal the arrival of the movement somatosensory feedback. We maintained the same PRE and POST nomenclature for the signals estimated at the cortical level.

All fMRI studies were performed on a 3.0-T scanner. Gradient echo EPI, sensitive to the BOLD effects, was performed using a commercial quadrature birdcage radio frequency coil. Subjects' heads were laid comfortably within the head coil and firm cushions were used to minimize the head motion. Forty axial (horizontal) slices with coverage of the whole brain were acquired. Acquisition parameters used in the functional scans were TE = 25 ms, TR = 3 s, flip angle = 908; 3 mm slice thickness with 0 mm gap; 64 × 64 acquisition matrix with a field of view (FOV) of 20 × 20 cm. The subjects were asked to perform the fast repetitive finger movement in the following sequence within one scan session: fixation (30 s), right-hand finger movements (30 s). The sequence was repeated five times during a single scan session. High-resolution structural images were also acquired in the axial plane (three-dimensional spoiled gradient recalled imaging [SPGR]). For two subjects a 1.9 mm thick contiguous slices were used, while for the other two subjects a slice thickness of 2 and 1.8 mm were employed.

We computed the statistically significant percentage increase of the fMRI signal during the task, compared to the resting state. Figure 10.4 shows the cortical connectivity patterns obtained for a representative subject in both experimental conditions (by use the EEG plus the fMRI information, (a) and (c), and the EEG information, (b) and (d)). Figure 10.4 (a) and (b) is related to the cortical connectivity patterns

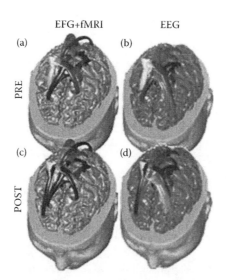

FIGURE 10.4 **(See color insert.)** (a) and (b): cortical connectivity patterns estimated before the EMG onset (PRE) in the finger-tapping condition, while (c) and (d) are related to the connectivity patterns computed after the EMG onset. Connectivity was estimated by DTF. (a) and (c): DTF estimated on the cortical activity reconstructed by means of the neuroelectric information (EEG only). (b) and (d): cortical connectivity patterns estimated with the use of multimodal integration of neuroelectric and hemodynamic data (EEG+fMRI). Same conventions of Figure 10.2.

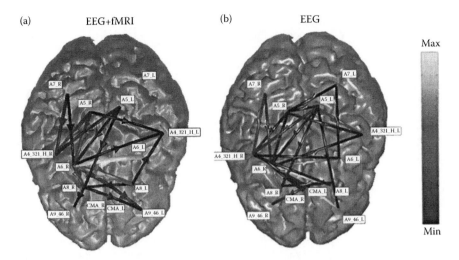

FIGURE 10.5 (**See color insert.**) Presents two cortical connectivity patterns obtained with the DTF estimator, in the beta band, during the congruent condition in the Stroop task. (a): cortical connectivity pattern is estimated with the use of the EEG and fMRI information; (b): the connectivity pattern is estimated from EEG information only.

estimated before the EMG onset (PRE), while Figure 10.4 (c) and (d) is related to the connectivity patterns computed after the EMG onset (POST). Details can be found in Babiloni et al. (2005) and Astolfi et al. (2005).

All statistical connections presented are significant at $p < 0.01$. A substantial agreement of the information conveyed by the two series of connectivity patterns is evident, although results for the EEG+fMRI case are more focused on a restricted circuit including the areas which were more active in the fMRI results. Changes in the intensity of the cortical connections are also evident in the case of the patterns estimated with the use of the EEG only, when compared to the connectivity patterns estimated with the use of EEG and fMRI. While the parietal and frontal connections are returned by both estimations, a shift of the intensity is observed in the connectivity patterns computed by using EEG and fMRI.

Similar results were obtained with the Stroop data, by estimating the cortical connectivity patterns in the beta band for the same subject with and without the fMRI information. Figure 10.5 shows two cortical connectivity patterns obtained with the DTF estimator in the beta band during the congruent condition of the Stroop task. In panel (a) of Figure 10.5, the cortical connectivity pattern is estimated with the use of the EEG and fMRI information; in panel (b), the connectivity pattern is estimated by the DTF applied to EEG data. As in the previous example, substantial agreement exists between the connectivity patterns obtained in the two conditions, which show the involvement of the parietal and frontal areas. This finding is similar to that already observed in the finger-tapping experiment, and also in this case, the intensity of the DTF estimated on the cortical waveforms obtained with the multimodal integration was higher than the one obtained by using the EEG information alone. Differences in the cortical pattern in different cortical areas could be noted, with the EEG+fMRI circuit being more focused and stronger.

10.4.3   EFFECT OF THE CHOICE OF THE MODEL ORDER

Computation of the connectivity pattern by DTF, PDC, and dDTF is based on the estimations of a multivariate autoregressive model (MVAR), as seen from the previous chapters. Since the accuracy of the MVAR modeling relies strongly on the determination of an appropriate model order p, it may be argued that the connectivity pattern estimation via DTF/PDC/dDTF can be strongly dependent from the choice of the model order. The optimal order had to be determined for each subject, as well as for each of the time segments considered in the estimated cortical waveforms. The Akaike information criteria (AIC) procedure is the most commonly used, even if, in some cases, the AIC curve as a function of the model order can reach a minimum value for more than one value of p. Therefore, it is interesting to investigate if slight changes in the choice of the model order can significantly affect the connectivity and

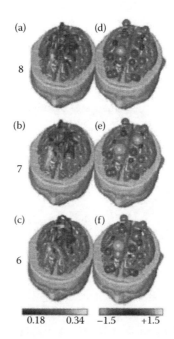

FIGURE 10.6   **(See color insert.)** Panels from (a) through (c): Connectivity patterns related to the different model orders used for the MVAR of the finger tapping data, starting from (c) the order 6 to (a) the order 8. Panels from (d) through (f): Inflow (red hues) and outflow (blue hues) patterns related to the same range of MVAR model orders. The computed connectivity patterns show a substantial similarity across different values of the model order in an interval around the optimal value of 7.

information flow patterns of the EEG data. In order to test this issue, we show a raw sensitivity analysis with different model orders, ranging from 6 to 8, on the estimated connectivity and outflow patterns of the finger-tapping EEG data discussed in the previous paragraph. Figure 10.6 shows the results obtained for the theta frequency band and the POST condition for testing the order from 6 to 8. A substantial equivalence of the estimated connectivity patterns computed with the different model orders can be seen from Figure 10.6. Similar results were found for the other frequency bands and the PRE time period (not shown here).

10.5   CONCLUSION

In this chapter, different methodologies for the estimation of cortical connectivity from neuroelectric and hemodynamic measurements have been reviewed and applied to a common data set in order to highlight the similarities and differences in the obtained results. On the basis of the results obtained in the previous paragraphs, we can conclude that (1) the different functional connectivity estimators (DTF, PDC, dDTF) return essentially the same global picture of connectivity patterns for the cognitive task investigated in a group of normal subjects. This was observed in the connectivity pattern analysis and, more specifically, for the indexes (as the outflow or the inflow) derived from such connectivity patterns; (2) The inclusion of spatial information from fMRI in the linear inverse estimation produces, from the same data set, connectivity patterns more focused and stronger with respect to those obtained only from EEG data; (3) The ROI approach is important in the estimation of cortical connectivity, since the distributed source models normally employed thousands of current equivalent dipoles, and for each one of them a cortical waveform is estimated. Applying the connectivity estimation directly to such cortical waveforms means to take into account thousands of time series, which is not currently possible. Apart from numerical and practical considerations, the identification processes for these large sets of variables would in fact require an extremely large amount of data points, exceeding the standard lengths of recordings actually performed. All these considerations suggest the necessity of the ROIs approach in the estimation of the cortical connectivity; and (4) The choice of the optimal model order p, though important for modeling and thus for the correct estimation of causality links, does not affect in a crucial way the connectivity estimation, providing that an average of the estimators is performed in frequency bands of interest and that the selected model order is in an interval around the optimal value.

In conclusion, in this chapter, we presented an integrated approach to estimate brain cortical connectivity information by using noninvasive methodologies involving the multimodal integration of electrophysiological measurements. These methodologies, starting from totally noninvasive recording techniques, allow the description of statistically significant cortical functional connections between different ROIs during simple motor or cognitive tasks in humans.

The toolbox employed for the estimation of DTF used in this chapter has been inserted in the freely available software eConnectome (He et al., 2011), at the following Internet address: http://econnectome.umn.edu.

REFERENCES

Astolfi, L., F. Cincotti, D. Mattia et al. 2005. Assessing cortical functional connectivity by linear inverse estimation and directed transfer function: Simulations and application to real data. *Clin. Neurophysiol. 116*: 920–932.

Astolfi, L., F. Cincotti, D. Mattia et al. 2006a. Comparison of different cortical connectivity estimators for high-resolution EEG recordings. *Hum. Brain Mapp. 28*: 143–157.

Astolfi, L., F. Cincotti, D. Mattia et al. 2006b. Assessing cortical functional connectivity by partial directed coherence: Simulations and application to real data. *IEEE Trans. Biomed. Eng. 53*: 1802–1812.

Astolfi, L., F. De Vico Fallani, F. Cincotti et al. 2007. Imaging functional brain connectivity patterns from high-resolution EEG and fMRI via graph theory. *Psychophysiology 44*: 880–893.

Astolfi, L., F. De Vico Fallani, F. Cincotti et al. 2009. Estimation of effective and functional cortical connectivity from neuroelectric and hemodynamic recordings. *IEEE Trans. Neural Syst. Rehab. Eng. 17*: 224–233.

Babiloni, F., C. Babiloni, F. Carducci et al. 2003. Multimodal integration of high-resolution EEG and functional magnetic resonance imaging data: A simulation study. *NeuroImage 19*: 1–15.

Babiloni, F., C. Babiloni, F. Carducci et al. 2004. Multimodal integration of EEG and MEG data: A simulation study with variable signal-to-noise ratio and number of sensors. *Hum. Brain Mapp. 22*: 52–62.

Babiloni, F., C. Babiloni, L. Locche et al. 2000. High-resolution electro-encephalogram: Source estimates of Laplacian-transformed somatosensory-evoked potentials using a realistic subject head model constructed from magnetic resonance images. *Med. Biol. Engin. Comp. 38*: 512–519.

Babiloni, F., F. Cincotti, C. Babiloni et al. 2005. Estimation of the cortical functional connectivity with the multimodal integration of high-resolution EEG and fMRI data by directed transfer function. *NeuroImage 24*: 118–131.

Baccalá, L. A. and K. Sameshima. 2001. Partial directed coherence: A new concept in neural structure determination. *Biol. Cybern. 84*: 463–474.

Dale, A., A. Liu, B. Fischl et al. 2000. Dynamic statistical parametric mapping: Combining fMRI and MEG for high-resolution imaging of cortical activity. *Neuron 26*: 55–67.

Gevins, A. 1989. Dynamic functional topography of cognitive task. *Brain Topogr. 2*: 37–56.

Granger, C. W. J. 1969. Investigating causal relations by econometric models and cross-spectral methods. *Econometrica 37*: 424–428.

de Peralta Menendez, R. G. and S. L. Gonzalez Andino. 1999. Distributed source models: Standard solutions and new developments. In *Analysis of Neurophysiological Brain Functioning*, ed. C. Uhl, 176–201. Heidelberg: Springer-Verlag.

Grinvald, A., E. Lieke, R. D. Frostig et al. 1986. Functional architecture of cortex revealed by optical imaging of intrinsic signals. *Nature 324*: 361–364.

Hansen, P. C. 1992a. Analysis of discrete ill-posed problems by means of the L-curve. *SIAM Rev. 34*: 561–580.

Hansen, P. C. 1992b. Numerical tools for the analysis and solution of Fredholm integral equations of the first kind. *Inverse Probl. 8*: 849–872.

He, B. and J. Lian. 2002. Spatio-temporal functional neuroimaging of brain electric activity. *Crit. Rev. Biomed. Eng. 30*: 283–306.

He, B., J. Hori, and F. Babiloni, 2006 EEG inverse problems. In *Wiley Encyclopedia in Biomedical Engineering*, Vol. 2, ed. M. Akay, 1355–1363. New York: Wiley.

He, B., Y. Dai, L. Astolfi et al. 2011. *econnectome*: A Matlab toolbox for mapping and imaging of brain functional connectivity. *J. Neurosci. Methods 195*: 261–269.

Im, C. H., Z. M. Liu, N. Zhang et al. 2006. Functional cortical source imaging from simultaneously recorded ERP and fMRI. *J. Neurosci. Methods 157*: 118–123.

Kamiński, M. and K. Blinowska. 1991. A new method of the description of the information flow in the brain structures. *Biol. Cybern. 65*: 203–210.

Kamiński, M., M. Ding, W. A. Truccolo et al. 2001. Evaluating causal relations in neural systems: Granger causality, directed transfer function and statistical assessment of significance. *Biol. Cybern. 85*: 145–157.

Liu, A.K., J. W. Belliveau, and A. M. Dale. 1998. Spatiotemporal imaging of human brain activity using functional MRI constrained magnetoencephalography data: Monte Carlo simulations. *Proc. Natl. Acad. Sci. USA 95*: 8945–8950.

Magistretti, P. J., L. Pellerin, D. L. Rothman et al. 1999. Energy on demand. *Science 283*: 496–497.

Nunez, P. L. 1981. *Electric Fields of the Brain*. New York: Oxford University Press.

Nunez, P. L. 1995. *Neocortical Dynamics and Human EEG Rhythms*. New York: Oxford University Press.

Puce, A., T. Allison, S. S. Spencer et al. 1997. Comparison of cortical activation evoked by faces measured by intracranial field potentials and functional MRI: Two case studies *Hum. Brain Mapp. 5*: 298–305.

Schelter, B., M. Winterhalder, M. Eichler et al. 2006. Testing for directed influences among neural signals using partial directed coherence. *J. Neurosci. Methods 152*: 210–219.

Stroop, J. R. 1935. Studies of interference in serial verbal reactions. *J. Exp. Psychol. 18*: 643–662.

Takahashi, D. Y., L. A. Baccalá, and K. Sameshima. 2007. Connectivity inference between neural structures via partial directed coherence. *J. Appl. Statist. 34*: 1259–1273.

Theiler, J., S. Eubank, A. Longtin et al. 1992. Testing for nonlinearity in time series: The method of surrogate data. *Physica D 58*: 77–94.

Urbano, A., C. Babiloni, P. Onorati et al. 1998. Dynamic functional coupling of high resolution EEG potentials related to unilateral internally triggered one-digit movements. *Electroencephalogr. Clin. Neurophysiol. 106*: 477–487.

Urbano, A., C. Babiloni, P. Onorati et al. 2001. Responses of human primary sensorimotor and supplementary motor areas to internally triggered unilateral and simultaneous bilateral one-digit movements. A high-resolution EEG study. *Eur. J. Neurosci. 10*: 765–770.

11 Methods for Connectivity Analysis in fMRI

João Ricardo Sato, Philip J. A. Dean, and Gilson Vieira

CONTENTS

11.1 INTRODUCTION

Functional magnetic resonance imaging (fMRI) has become one of the most prominent neuroimaging tools for the *in vivo* study of brain function. Since the BOLD effect was described by Ogawa et al. (1990), the number of studies involving fMRI has increased exponentially. It is a noninvasive technique with high spatial resolution, and these factors have contributed greatly to its success. MRI does not require the use of radiation or radioactive isotopes, as is the case in computed tomography (CT) and positron emission tomography (PET). As such, the risk factors associated with the protocol are significantly smaller, allowing more comprehensive use within both clinical and nonclinical populations. Acquisition of data within fMRI is based on the paramagnetic properties of deoxyhemoglobin and hemodynamic coupling, producing the BOLD contrast. Logothetis et al. (2001) revealed that the BOLD signal can be considered as an indirect measure of local synaptic activity, as it shows greater correlation with local potentials than multi-unit activity. The spatial resolution of fMRI is greater than both PET and EEG, allowing finer scale analysis of this local synaptic activity. In addition, an MRI scanner allows a variety of neurological data to be obtained, from fMRI to structural (T1/T2 images; diffusion tensor imaging, DTI) and even metabolic (magnetic resonance spectroscopy, MRS) data sets. As such, its utility in a research and clinical setting is growing. This is evidenced by several hospitals now acquiring MRI systems with magnetic fields greater than 1.5, as well as software and hardware capable of acquiring fMRI data, at least for resting state protocols.

During the first decade after the description of the BOLD effect, most fMRI studies concentrated on the identification of brain regions activated during certain tasks or by specific stimuli. Historically, the majority of functional mapping in cognitive science was based on documenting the behavior of patients with lesions, and relating the changes in behavior to the damaged brain areas. With the advent of fMRI, researchers were now able to examine and localize brain function in healthy individuals, and at a greater spatial resolution. Although fMRI could be said to have initiated a revolution in neuroscience, and led to substantial innovations in the related literature, some scientists were afraid that it merely represented a "new phrenology."

There was some substance to these fears, since most studies at the time were limited to mapping stimuli to brain regions. The simple identification of activated regions is relevant to locate "where" the information is processed (functional specialization), but it does not give any information about "how" this processing occurs or how each region affects each other (functional integration). It is well established that the structural and functional organization of the brain is based on a complex network of interdependent nodes. Furthermore, several cognitive models are influenced by connectionist theories, for which independent activation of separated regions are not enough to produce the requisite behavior.

To overcome these limitations, there was an increasing interest in the exploration of new approaches using fMRI aimed at identifying regions with synchronous neural activity and the information flow between these regions. Biswal et al. (1995) were one of the pioneers in this field. They demonstrated that there was a strong synchronization between BOLD signals within the motor system, even during a resting state paradigm. Friston (1994) defined two terms to describe the functional relationship between brain regions: functional connectivity and effective connectivity. Functional connectivity is used to describe correlations between neural activity generated by spatially remote brain areas, while effective connectivity refers to the causal influences that these regions exert over each other. In others words, functional connectivity analysis is focused on detecting common modulations in neural activity between different regions, while effective connectivity analysis tries to uncover the direction that information flows within this network. Since the time of their definition, studies which examine connectivity and functional integration using fMRI have been classified into these two categories. Functional connectivity analysis is more practical in situations when there is little *a priori* information about the network of regions involved in a specific paradigm, as this limits the hypotheses that can be tested. This results in a more open-ended investigation of connectivity patterns. However, if there is enough information about the network structure to be modeled, effective connectivity is a much more powerful tool to test hypotheses and explain data.

In the context of psychiatric and neurological disorders, connectivity analyses have become a promising tool with which to explore altered brain functioning in patients. Rowe (2010) describes several studies which demonstrate that connectivity analyses can be more sensitive to diseases than conventional analysis of regional differences in activation. This suggests that the investigation of the connectivity within distributed brain networks can significantly enhance the comprehension of brain disorders.

In this chapter, we discuss the most popular methods for functional and effective connectivity analysis of fMRI data. The foundations, properties, advantages, and pitfalls are also described. In addition, we include several illustrative examples based on real fMRI acquisitions to highlight the specific procedures and outcomes of each approach.

11.2 DATA STRUCTURE AND PREPROCESSING

During an fMRI experiment, several echo-planar imaging (EPI) volumes are acquired over time, while the subject performs certain tasks or remains at rest. The MRI scanner acquires the volumes in a slice-by-slice fashion and a volume is a set of these slices. Each slice has a thickness which depends on the scanning protocol, and within each slice, data are collected at a number of locations. These locations represent the smallest volumetric units of the MRI data set, and they are termed voxels. The EPI volumes are stored in digital format as a matrix where each cell contains the BOLD signal intensity measured at one particular voxel within one particular slice. As more volumes are obtained, the data represent the fluctuations in BOLD signal intensity at each voxel over the course of data acquisition.

The images are usually preprocessed using head motion correction, spatial normalization, and spatial smoothing prior to any statistical analysis. Therefore, the procedures for brain activation mapping and connectivity analysis are carried out after pre-processing. Some additional optional steps are slice-timing correction, temporal filtering, B0 unwarping, regression on head motion parameters, and/or GM, WM, CSF, or global signal. More details about fMRI data pre-processing can be found in Jezzard et al. (2003) and Penny et al. (2006). These methods can be implemeted using freely available fMRI analysis packages such as FSL* and SPM.[†]

In this chapter, the illustrative data sets were preprocessed using FSL, and all statistical analysis were performed using R platform,[‡] and the libraries AnalyzeFMRI,[§]

---

* http://www.fmrib.ox.ac.uk/fsl/
[†] http://www.fil.ion.ucl.ac.uk/spm/
[‡] http://www.r-project.org
[§] Coded by J. L. Marchini and P. Lafaye de Micheaux.

SEM,[*] and FIAR[†]. In R programming, the set of preprocessed EPI volumes from a single run is stored as a 4D array structure (x, y, z and time), which we will refer to as **volume**. Thus, **volume[x, y, z, t]** refers to the BOLD signal at time t measured at voxel at coordinates x, y, and z. For didactical purposes, the analyses are shown in their simplest version. Several improvements can be included in order to increase the sensitivity of the methods, but these improvements are not within the scope of this chapter.

In addition, multiple comparisons correction is frequently applied to brain mapping procedures, mainly at the group analysis level. There are several approaches (e.g., FWE and FDR), which are applied at different spatial levels (voxel and cluster) to deal with these problems. However, since the examples shown are based on individual data and are for visualization and comparison purposes, the maps we present in this chapter are not corrected for multiple comparisons. Further information about multiple comparison problems in neuroimaging and the methods used to tackle these problems can be found in Penny et al. (2006).

11.3 ILLUSTRATIVE DATA SETS

One of the purposes of this chapter is to illustrate how connectivity analyses can be applied to real fMRI data, and the results this can produce. The codes and acquisition files used are enclosed with this book. The two fMRI acquisitions were approved by the local ethics committee.

Resting State Data Set: one participant was asked to lie in the MRI scanner with eyes closed for a resting state fMRI session. The volumes were acquired using a Siemens 3 Tesla MR System (Erlangen, Germany) with a T2*-weighted EPI sequence (TR=1 800 ms, TE=30 ms, 64 × 64 matrix, FOV = 240 × 240 mm, inter-slice gap = 0.6 mm, voxel size = 3.75 × 3.75 × 6.00 mm, 18 slices covering the whole brain). A total number of 450 volumes were acquired.

Stroop Test Data Set: one participant was asked to undergo a Stroop Test (Peterson et al., 1999) inside the scanner. This task consisted of the visual presentation of words written in three different colors (red, blue, and green). The participant is required to respond to the color of the word, not its meaning. If the color of the word is the same as its meaning (congruent; e.g., the word blue, colored blue), participants respond faster than if the color of the word is different from its meaning (incongruent; e.g. the word blue, colored green). There is also a third, neutral, condition, whereby words with meaning unrelated to color are presented. The participant responded to each trial using their right hand by pressing one of three response alternatives (red, blue, or green) on an MRI-compatible button box. The experiment used a block design, with each block composed of 10 trials (corresponding to 10 EPI volumes). The sequence of conditions was congruent–neutral–incongruent, repeated six times. The volumes were acquired using a Siemens 3 Tesla MR system (Erlangen, Germany) with a T2-weighted EPI sequence (TR = 2000 ms, TE = 50 ms, 64 × 64 matrix, FOV = 200 ×

---

[*] By John Fox.
[†] By Bjorn Roesltraete.

200 mm, inter-slice gap $= 0.5$ mm, voxel sixe $= 3 \times 3 \times 3$ mm, 32 slices covering the whole brain). A total number of 180 volumes were acquired.

11.4 FUNCTIONAL CONNECTIVITY ANALYSIS

One of the most attractive properties of functional connectivity analysis is its simplicity. It is one of the most suitable approaches for exploratory studies, as it does not require any strong assumptions. This also aids its use in clinical studies, which usually research the effect of disease or damage on brain function. Functional connectivity analysis can usually be carried out on these data, whereas the greater assumptions and constraints needed for effective connectivity analysis mean that brain damage and disease often reduce its efficacy. Functional connectivity analysis can therefore provide new insights on brain function or disease state by means of exploratory analysis. However, the interpretation of results from these analyses is limited by their simplicity, since they only refer to common fluctuations of neural activity between different brain regions. This does not necessarily mean that these regions are directly anatomically connected, or that there is a causal relationship in their mutual variations in activity.

11.4.1 PEARSON'S CORRELATION MAPPING

Pearson's correlation mapping was the first approach employed to study functional connectivity, and is still one of the most widely used. This approach produces a Pearson correlation coefficient (PCC) for BOLD signals measured at different regions, which allows for simplicity of analysis and results interpretation that may explain its popularity. Essentially, the PCC measures the linear dependence between two variables (BOLD signals in this case) on a scale from -1 to 1, and is zero if the two variables are linearly unrelated.

   Correlation maps are obtained by following these steps:

Step 1: Define a seed voxel of interest.
Step 2: Calculate the Pearson correlation between the BOLD signal of the seed and all other voxels in the brain.
Step 3: Apply a threshold to this correlation map.

   These maps illustrate the regions in the brain where the BOLD signal has a similar (positive or negative correlation) pattern of variation to the signal measured at the seed voxel. They are not equivalent to the conventional activation maps, since the activation in two regions can be correlated without being entirely related to the experimental design or stimuli. It is important to note that many studies define a number of voxels in a region-of-interest (ROI) as the seed, instead of a single voxel. This increases the signal-to-noise ratio at the seed and reduces the influence of problems associated with spatial normalization. In this case, the average or first principal component for all voxels within the ROI is assumed to be the regional representative.

There are other variations on correlation mapping that can be used on specific data types, or for specific analyses. Spearman's correlation coefficient is a useful alternative to PCC when the data contain nonlinear monotonic relationships or outliers. Common confounding variables within fMRI data, such as cardiac and respiratory rhythms, are related to the hemodynamic coupling on which the image signal is based. These can affect functional connectivity mapping as they produce alterations in the BOLD response measured by fMRI. If data on these rhythms are acquired during scanning, it is very common to partialize the signals by these confounder variables, or even analyze the BOLD signal of voxels from an ROI at a blood vessel as a control. In addition, multiple regression analysis is frequently applied when researchers need to address confounding variables within their data.

We will now present an illustrative example of this type of analysis using the Resting State data set. The seed voxel is placed at the left primary motor cortex (M1), at coordinates (29, 26, 31 in voxels). The R code for a voxel-by-voxel correlation analysis is given by

```
SEED=volume[29,26,31,]
MAP=array(0,c(dim(volume)[1:3],1))
for(xi in 1:dim(volume)[1]){
   for(yi in 1:dim(volume)[2]){
    for(zi in 1:dim(volume)[3]){
     MAP[xi,yi,zi,1]=cor(SEED,volume[xi,yi,zi,])
    }
   }
}
```

where the correlation map is stored in the object **MAP**. The resulting correlation map is shown in Figure 11.1.

In accordance with the results found by Biswal et al. (1995), there was a strong synchronization between BOLD signals within the motor system during this resting state

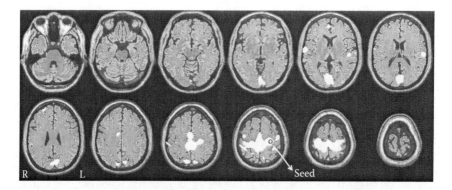

FIGURE 11.1 **(See color insert.)** Correlation map for resting state data set ($r > 0.65$). The seed was placed at M1 in the left hemisphere (Medical image format inverts left and right, as can be seen by R and L above). Red indicates positive correlation. Note that the right hemisphere M1 and SMA are correlated to the seed voxel.

paradigm. The BOLD signals measured at the bilateral M1, bilateral premotor areas (PMA), and supplementary motor area (SMA) are strongly positively correlated. In addition to motor areas, primary visual areas (V1) and somatosensory areas (S1), as well as prefrontal areas (middle frontal gyrus and dorsolateral prefrontal cortex, DLPFC) are also positively correlated.

11.4.2 INDEPENDENT COMPONENT ANALYSIS

As described in the previous section, correlation mapping requires *a priori* specification of a seed voxel or ROI. An attractive alternative to this approach is to analyze the data at all intracranial voxels to identify temporally linked activation patterns, or components, across the brain. The components can then be used to identify regions which are active in parallel, and may therefore be functionally connected. Separating data into these components is often compared to the "cocktail party problem." In this situation, there is a cocktail party where several people are talking at the same time, and several microphones are distributed around the room. Each microphone records the voices of many different people in the room, so that all recorded signals will consist of a mixture of different voices. In order to find what each individual was saying, we must separate the voice of each person (the component) from the mixture of voices recorded by the microphones. This separation of mixed signal data into its components is called blind source separation, and one of the most popular approaches to deal with it is independent component analysis (ICA; Comon, 1994; Hyvörinen, 1999). The principle of ICA is the estimation of those components in the data which are maximally independent from each other. ICA was first introduced to the analysis of fMRI data by McKeown et al. (1998) and since then, it has been extensively applied as an exploratory analysis tool, predominately in resting state data acquisition. As there is no seed region to be defined, ICA can even be considered more exploratory than correlation mapping. ICA does not require or estimate any anatomical connectivity or directionality of information flow, but simply identifies common components distributed in several brain regions.

The ICA model assumes that the signals measured at time t at voxels v_1 to v_n are linear combinations of independent components c_1 to c_k and a random error u_i $(i = 1, \ldots, n)$:

$$v_{1,t} = A_{1,1}c_{1,t} + A_{1,2}c_{2,t} + \cdots + A_{1,k}c_{k,t} + u_{1,t}$$

$$\vdots$$

$$v_{n,t} = A_{n,1}c_{1,t} + A_{n,2}c_{2,t} + \cdots + A_{n,k}c_{k,t} + u_{n,t}.$$

The estimation procedures focus on obtaining the independent components c_1 to c_k and the mixing coefficients $A_{i,j}$ $(i = 1, \ldots, n; j = 1, \ldots, k)$ while minimizing mutual information or maximizing non-Gaussianity. Using this method, the ICA-map of component j is the brain map composed of the coefficients $A_{i,j}$. In addition, these coefficients can also be transformed to a z-statistic for thresholding and visualization.

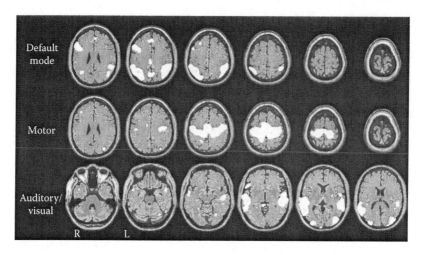

FIGURE 11.2 (See color insert.) ICA mapping for resting state data set, illustrating components 3 (default mode network, top row), 11 (motor network, middle row), and 17 (auditory/visual network, bottom row) ($|z| > 2.00$). Red and blue indicate positive and negative coefficients, respectively.

Further details about ICA estimation and modeling can be found in Hyvörinen (1999) and Beckmann and Smith (2004).

The R code for ICA analysis of the Resting State data is given by

```
Ncomp=30   #Number of components to be extracted
FIT=f.ica.fmri(FILENAME, Ncomp, norm.col=TRUE,alg.type="deflation")
```

This analysis yielded the components shown in Figure 11.2 (components 3, 11, and 17, which represent the default mode, motor, and auditory/visual networks, respectively). The three components shown reveal easily distinguishable connectivity networks, which are well established from previous literature concerning resting state networks (van den Heuvel and Hulshoff Pol, 2010).

The default mode network is thought to represent internally focused tasks such as autobiographical and episodic memory recall and consolidation. The ICA analysis has identified a default mode network component with common activations in the posterior cingulate cortex, inferior parietal lobule, and Broca's area, in keeping with previous studies (Buckner et al., 2008), as well as activation in the dorsal anterior cingulate cortex (ACC). The motor component is similar to that found using Pearson's correlation mapping, in that bilateral M1, bilateral PMA, and SMA are concurrently activated. Finally, the component representing the audio-visual network demonstrates bilateral common fluctuation of a large area around the middle temporal gyrus (an area involved in auditory processing and language, comprising Brodmann areas 21 and 22) and bilateral deactivation of primary and secondary visual areas (V1 and V2).

These results demonstrate the utility of the ICA method in revealing possible common fluctuations for further investigation. However, these common fluctuations do not necessarily constitute causal connections, and there may be better methods with which to investigate, for example, the connectivity of auditory processing in more detail.

11.4.3   Psychophysiological Interactions

The psychophysiological interactions (PPI) approach was first proposed by Friston et al. (1997) as an extension of conventional correlation mapping for use on data where it is believed that the functional connectivity between regions is being modulated by the experimental task or stimuli. Thus, the aim of PPI is to find intracranial voxels whose correlation with the seed voxel (or ROI) differs depending on the experimental condition.

The basic idea of PPI is the following: suppose that the variable x_t $(t = 1, \ldots, n)$, where n is the total number of EPI volumes, contains the expected hemodynamic response at time t (box-car or convolved response). The activation of a voxel at time t can be modeled by

$$\text{voxel}_t = a + b * x_t + u_t,$$

where a and b are scalars, and u_t is a random noise with mean zero. Thus, by adding the seed BOLD signal to the model, we have the PPI model:

$$\text{voxel}_t = a + b * x_t + c * \text{seed}_t + p * x_t * \text{seed}_t + u_t,$$

where p is a scalar, related to the interaction term $x_t * \text{seed}_t$, which measures the changes in functional connectivity between the seed and the voxel of interest for the different experimental conditions which are implicit in x_t. PPI maps depict a voxel-by-voxel analysis of this coefficient, p, or of its statistical significance.

The R code for PPI analysis of the Stroop Test data set, comparing the congruent and incongruent conditions (after removing the volumes under neutral condition from the data), is given by

```
PPIconvVSincong=array(0,c(dim(volume)[1:3],1))
for(xi in 1:dim(volume)[1]){
  for(yi in 1:dim(volume)[2]){
    for(zi in 1:dim(volume)[3]){
      FIT=lm(volume[xi,yi,zi,]~incong+SEED+SEED*incong)
      PPIcongVSincong[xi,yi,zi,1]=qnorm(1-summary(FIT)$coef[4,4])
    }
  }
}
```

Note that, in this example, we convert the p-value of the coefficient p to a z-statistic. The PPI map for this analysis, which employs the ventral ACC as the seed, is shown in Figure 11.3.

As expected, the areas which vary their functional connectivity as a result of task difficulty (incongruent vs. congruent conditions) are those previously implicated in the performance of the Stroop task (Peterson et al., 1999; Zoccatelli et al., 2010). The medial prefrontal cortex (MPFC), bilateral DLPFC, SMA, and superior temporal

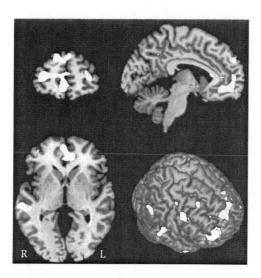

FIGURE 11.3 **(See color insert.)** PPI analysis mapping for Stroop's task data set, with the seed placed at the ventral ACC ($|z| > 3.00$). Red indicates increased connectivity as an effect of increasing task difficulty (incongruent>congruent).

gyrus have all been shown to interact with the ACC to aid the conflict monitoring and response making required by the task.

In order to demonstrate how the PPI data are formed from individual BOLD signals, a scatter-plot illustrating the correlation between PPI at the ACC and the MPFC is shown in Figure 11.4. The correlation between these two regions is significantly greater during the incongruent condition when compared to the congruent one, as expected from the results shown in Figure 11.3.

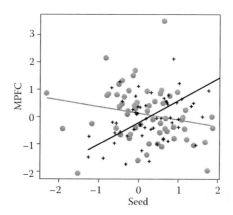

FIGURE 11.4 Scatter plot showing correlation between the BOLD signal at the seed (ventral ACC) and the medial prefrontal cortex (MPFC), during the congruent (gray circles) and incongruent (black crosses) conditions of the Stroop Test. Fitted lines illustrate the direction and polarity of this correlation.

11.4.4 GRANGER CAUSALITY MAPPING

One of the major limitations of correlation-based functional connectivity mapping is that correlation measures are symmetrical. In other words, the correlation between the variables A and B is exactly the same as that between B and A. This means that the correlation coefficient can identify linear relationships and association between the signals of two brain regions, but cannot provide any insight into the direction of this information flow. As described in other chapters from this book, identification of Granger causality (Granger, 1969) using vector autoregressive (VAR) models could potentially provide inferences about the temporal precedence and direction of information propagation between brain regions. Goebel et al. (2003) applied Granger causality to imaging data to create Granger causality mapping (GCM), which can be seen as an extension of conventional correlation mapping. The VAR model for a seed and a voxel of interest is given by

$$\text{seed}_t = a_1 + b_1 \text{seed}_{t-1} + \cdots + b_p \text{seed}_{t-p}$$
$$+ c_1 \text{voxel}_{t-1} + \cdots + c_p \text{voxel}_{t-p} + u_{1t}$$
$$\text{voxel}_t = a_2 + d_1 \text{seed}_{t-1} + \cdots + d_p \text{seed}_{t-p}$$
$$+ e_1 \text{voxel}_{t-1} + \cdots + e_p \text{voxel}_{t-p} + u_{2t},$$

where the coefficient c_i $(i = 1, \ldots, p)$ refers to the information flow from the voxels of interest to the seed, and the coefficient d_i $(i = 1, \ldots, p)$ refers to the information flow from the seed to the voxel. The coefficients a_1 and a_2 are intercepts and the coefficients b_i and e_i are autoregressive terms. Granger causality analysis for the influence of the seed on the voxel tests whether there is at least one coefficient d_i that differs from zero. In a similar fashion, the Granger causality from the voxel to the seed tests whether there is at least one coefficient c_i that differs from zero.

The main steps of this procedure are:

Step 1: Define a seed voxel of interest.

Step 2: Estimate a bivariate VAR model for the BOLD signal of the seed and all other voxels in the brain.

Step 3: Calculate the *F*-statistics for the seed *Granger causing* the other voxels. For visualization or group analysis purposes, these *F*-statistics can be converted to *z*-statistics.

Step 4: Apply a threshold to this seed-sends map, considering the statistical significance at each voxel.

Step 5: Calculate the *F*-statistics for the other voxels *Granger causing* the seed. For visualization or group analysis purposes, these *F*-statistics can be converted to *z*-statistics.

Step 6: Apply a threshold to this seed-receives map, considering the statistical significance at each voxel.

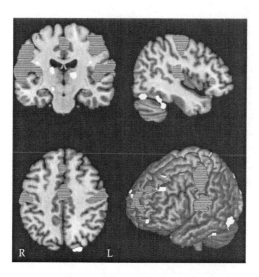

FIGURE 11.5   (**See color insert.**) Granger causality maps for the resting state data, with the seed placed at the left hemisphere M1 ($|z| > 7.00$). The voxels which send information to the seed are depicted in blue, whereas those that receive information from the seed are depicted in green.

The R code for voxel-by-voxel Granger causality mapping is very similar to the one for correlation mapping, replacing the correlation calculation by the statistics of step 5 (for more information, see accompanying code files). An illustrative example of Granger causality mapping applied to the resting state data set is depicted in Figure 11.5. The seed voxel is placed at left the hemisphere M1 at the coordinates (29, 26, 31 in voxels).

These data build on the results of the previous Pearson's correlation mapping to show the directionality of information flow in the motor system during a resting state. Bilateral M1, bilateral PMA, and SMA all send information to the left hemisphere M1. In turn, the left hemisphere M1 sends information to the left cerebellum, bilateral thalamus, and a variety of frontal areas and white matter tracts. It is known that there are bilateral connections between the primary motor cortices, as well as between these primary areas and secondary motor areas such as PMA and SMA. This network is used in the coordination and planning of movement. That the left hemisphere M1 sends information to the cerebellum is in keeping with this network for complex movements. Information sent to the thalamus, white matter tracts, and frontal areas suggests that the left M1 is feeding back its activation state to a variety of areas within the brain to allow functional integration.

11.5   EFFECTIVE CONNECTIVITY ANALYSIS

The main benefit of effective connectivity analysis is the ability to describe connectivity in terms of the propagation and direction of flow of information through the different brain regions. Quantification of the causal influences between the brain regions, selected on the basis of a theoretical cognitive model, may provide significant insights about brain function. As previously stated, the main obstacles to performing a reliable effective connectivity analysis are the amount of prior information about the process studied, which is needed to create the cognitive model, and the validity of the assumptions required by the specific method used. In this section, we describe the three most popular approaches to effective connectivity analysis. It is important to emphasize that these methods are complementary, since they extract different features from the data, and differ in the requirements for their application. No one method is always superior and each one may be more suitable than the others, depending on the data acquired and the process studied.

11.5.1 STRUCTURAL EQUATIONS MODELING

Structural Equation Modeling (SEM) is frequently applied as a statistical tool to quantify (or test) causal relationships between variables (Wright, 1921; Simon, 1953). In econometrics, it has been commonly applied to quantify simultaneous systems such as supply–demand models. SEM has also been used in psychometrics for modeling the interactions between different variables. In most cases, SEM is more suitable as a confirmatory analysis, mainly concerned with quantifying the strength of connectivity and testing specific hypotheses raised by the researcher. In the context of fMRI experiments, SEM was introduced by McIntosh (1998) to describe the effective connectivity in a memory task.

Essentially, SEMs are estimated using maximum likelihood and the statistical significance of each link can be assessed by likelihood ratio tests. Suppose we have the signals of four regions A, B, C, and D and an *a priori* model that A influences B, B influences C and D, and finally, C influences D. These causal relations can be visualized in a diagram, shown in Figure 11.6. This diagram can be represented in a simultaneous equation system given by

$$\begin{cases} A_t = s_A u_{At} \\ B_t = b_A A_t + s_B u_{Bt} \\ C_t = c_b B_t + s_C u_{Ct} \\ D_t = d_B B_t + d_C C_t + s_D u_{Dt}, \end{cases}$$

where u_{At}, u_{Bt}, u_{Ct}, and u_{Dt} are random errors with standard Gaussian distribution and s_A, s_B, s_C, s_D, b_A, c_B, d_B, and d_C are coefficients to be estimated. The main assumptions when using SEM are that the prespecified model is correct and the nodes

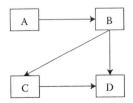

FIGURE 11.6   Illustrative example of a causal diagram.

are linearly dependent, that the random errors follow a Gaussian distribution, and that all relevant variables are included in the model. SEM uses the zero-lag covariance between regions only, which ignores the temporal structure of the data. This means that shuffling the data in time would have no influence on the estimation of SEM parameters. In cases where there are two different experimental conditions (e.g., resting and task), SEM can be fitted using the data points (volumes) of each condition.

SEM analysis of congruent and incongruent conditions of the Stroop Test data set was carried out as an illustrative application. The *a priori* model, adapted from previous findings by Schlösser et al. (2008), is shown in Figure 11.7a. It includes the rostral ACC (rACC), dorsal ACC, (dACC), ventrolateral pre-frontal cortex (VLPFC), DLPFC, and V1. The R code for this analysis is

```
#Load SEM library
require(sem)
#Specify Model
MODELO <- specify.model()
 dACC->rACC, dACC2rACC, NA
 rACC->dACC, rACC2dACC, NA
 DLPFC->dACC, DLPFC2dACC, NA
 VLPFC->dACC, VLPFC2dACC, NA
 dACC->DLPFC, dACC2DLPFC, NA
 V1->DLPFC, V12DLPFC, NA
 V1->VLPFC, V12VLPFC, NA
 dACC->VLPFC, dACC2VLPFC, NA
 VLPFC->DLPFC, VLPCF2DLPFC, NA
 dACC<->dACC, sigmadACC, NA
 rACC<->rACC, sigmarACC, NA
 DLPFC<->DLPFC, sigmaDLPFC, NA
 VLPFC<->VLPFC, sigmaVLPFC, NA
 V1<->V1, sigmaV1, NA

#Estimate SEM parameters
CORREL=cor(cbind(rACC,dACC,DLPFC,VLPFC,V1))
FIT $<$- sem(MODELO,S= CORREL, N=length(rACC))
summary(FIT)
```

The estimated parameters for each condition are shown in Table 11.1. Parameters that were statistically different from zero ($p < 0.05$) are depicted in Figure 11.7b as the estimated model. The results imply that the incongruent condition requires more interaction within the network, as relatively more causal relations were detected than in a congruent condition. However, direct subtraction of parameter estimates between the two conditions does not highlight any significant differences. A possible recommendation in this case would be to increase the sample size (i.e., the number of volumes acquired). It is likely that the current data, comprising one participant with 60 volumes per condition, do not have enough statistical power to find a significant effect of condition.

TABLE 11.1
SEM Analysis of the Stroop Task Data Set

	Connection	Estimate	Std. Error	*p*-value
Congruent	dACC to rACC	0.604	0.356	0.09
Condition	rACC to dACC	−0.341	0.424	0.421
	DLPFC to dACC	0.218	0.519	0.674
	VLPFC to dACC	0.399	0.297	0.179
	dACC to DLPFC	0.12	0.441	0.785
	V1 to DLPFC	−0.106	0.141	0.454
	V1 to VLPFC	0.595	0.118	**<0.001**
	dACC to VLPFC	−0.025	0.223	0.91
	VLPCF to DLPFC	0.486	0.252	0.054
Incongruent	dACC to rACC	0.98	0.235	**<0.001**
Condition	rACC to dACC	−0.971	0.522	0.063
	DLPFC to dACC	−0.011	0.34	0.974
	VLPFC to dACC	0.992	0.349	**0.004**
	dACC to DLPFC	0.413	0.208	**0.047**
	V1 to DLPFC	0.245	0.139	0.077
	V1 to VLPFC	0.581	0.125	**<0.001**
	dACC to VLPFC	−0.084	0.22	0.703
	VLPCF to DLPFC	−0.056	0.173	0.748

Note: Parameter estimates, standard errors, and significances are shown for each connection in the model (specified in Figure 11.7a) and for congruent and incongruent conditions. Parameters that reached significance are highlighted in bold font.

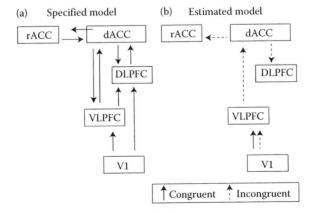

FIGURE 11.7 Prespecified connectivity model (a) and SEM estimated model (b) showing significant links (*p* < 0.05 compared to zero). The model includes the rostral anterior cingulate cortex (rACC), dorsal anterior cingulate cortex (dACC), dorsolateral prefrontal cortex (DLPFC), ventrolateral prefrontal cortex (VLPFC), and primary visual cortex (V1).

11.5.2 Dynamic Causal Models

The dynamical causal model (DCM) for fMRI was first introduced by Friston et al. (2003) and can be seen as an extension of SEM and PPI, focusing on the inherent difficulties of inferring neural interactions when observing hemodynamic signals during different tasks. According to the authors, a common limitation of all previous approaches to connectivity analysis is that they focus solely on the identification of relations between observed signals and not between the neural activities (which are latent variables) at different nodes. DCM attempts to estimate and model the parameters of the network structure, including the parameters of the biophysical process which links neural activity to hemodynamic coupling. The basic idea of DCM is to consider the brain as a nonlinear dynamic system, in which the inputs are deterministic and specified by the experimental design (stimulus) and the output is the BOLD signal measured using fMRI. Since the number of parameters is huge and the model is complex and nonlinear, estimation is carried out using the Bayesian inferences, based on the Monte Carlo Markov chain and expectation-maximization algorithms (Friston et al., 2003).

Basically, in the case of K regions of interest, DCM assumes that the derivatives (temporal changes) of neural activity z (a vector with dimension K) is a function of the local neural activity and of the stimuli input u:

$$\dot{z} = Az + \sum_{i=1}^{m} u_i B_i z + Cu,$$

where A is a $K \times K$ matrix of intrinsic connectivity, B_i is a $K \times K$ matrix of connections which are modulated by the stimuli (which have m active experimental conditions), and C is a vector of activity modulation by the stimuli. Note that DCM has specific parameters for the "baseline" connectivity, connections influenced by the experimental conditions, and the activation induced by the stimuli. DCM assumes that the BOLD signal measured within the scanner is related to the neural activity (z) as a result of a coupling process assuming the Balloon Model (Buxton et al., 2004). Thus, estimation of DCM involves obtaining estimates for A, B_i ($i = 1, \ldots, m$), C, and the parameters of the Balloon model.

The package FIAR (developed by Bjorn Roesltraete) contains routines for specifying and estimating DCM models within the R platform. An illustrative example of DCM analysis applied to the Stroop data set was performed. The DCM specification for the model represented in Figure 11.8 (based on results from Schlösser et al. (2008), modified from Figure 11.7) is given by

```
DCM=list()
DCM$n=5 #number of areas
DCM$m=2 #number of active conditions
DCM$ons$<$-list()
# ONSETS of congruent condition (in scans)
DCM$ons$input1$<$-c(0,30,60,90,120,150)
# ONSETS of incongruent condition (in scans)
DCM$ons$input2$<$-c(0,20,50,80,110,140,170)
```

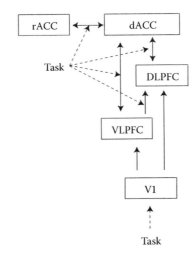

FIGURE 11.8 Prespecified connectivity model for DCM. The model includes the rostral anterior cingulate cortex (rACC), dorsal anterior cingulate cortex (dACC), dorsolateral prefrontal cortex (DLPFC), ventrolateral prefrontal cortex (VLPFC), and primary visual cortex (V1).

```
DCM$dur$<$-list()
DCM$dur$input1$<$-10 #durations of congruent block (in scans)
DCM$dur$input2$<$-10 #durations of incongruent block (in scans)
#intrinsic connections
DCM$a=c(0, .3, .3, .3, 0,
 .3, 0, 0, 0, 0,
 .3, 0, 0, .3, .3,
 .3, 0, 0, 0, .3,
 0, 0, 0, 0, 0)
#functional connections induced by congruent condition
DCM$b=c(0, .3, .3, .3, 0,
 .3, 0, 0, 0, 0,
 .3, 0, 0, .3, 0,
 .3, 0, 0, 0, 0,
 0, 0, 0, 0, 0,
#functional connections induced by incongruent condition
 0, .3, .3, .3, 0,
 .3, 0, 0, 0, 0,
 .3, 0, 0, .3, 0,
 .3, 0, 0, 0, 0,
 0, 0, 0, 0, 0)
# modulation of congruent condition on V1
 DCM$c=c(0, 0, 0, 0, .5,
# modulation of incongruent condition on V1
 0, 0, 0, 0, .5)
#hemodynamic parameters (these are standard values)
DCM$h=c(.65, .41, .98, .32, .34,0)
```

```
DCM$x $<$- 5*DCM$n # number of states
  (these are standard values)
DCM$HPF=0 #High-pass filter
#Scanner parameters
DCM$TR=2 #repetition time in seconds (integer)
DCM$TE=0.04 #echo time in seconds (integer)
DCM$v=180 #number of scans (integer)
DCM$T=16 #number of timebins (integer, this is a
  standard value)
```

Finally, the model can be estimated using the function spm.dcm.estimate within the package FIAR. The estimates for the matrices A, B, and C are shown in Table 11.2. The majority of numbers in the table are very small, possibly due to the small sample size (number of participants and number of volumes). However, DCM parameter estimates suggest an increase in the influence of DLPFC and VLPFC over dACC during the incongruent condition when compared to the congruent condition (as shown by the numbers highlighted in bold font).

TABLE 11.2
DCM Analysis, and Parameter Estimates, of the Stroop Task Data Set Using the Model Specified in Figure 11.8

Matrix	To\From	dACC	rACC	DLPFC	VLPFC	V1
A	dACC	—	2.06E−06	0.01373	0.02046	0
	rACC	0.00111	—	0	0	0
	DLPFC	0.00026	0	—	0.00436	0.06028
	VLPFC	0.00037	0	0	—	0.09138
	V1	0	0	0	0	—
B	dACC	—	2.05E−06	**0.01507**	**0.02247**	0
Incongruent	rACC	0.00128	—	0	0	0
	DLPFC	0.0003	0	—	0.00664	0
	VLPFC	0.00033	0	0	—	0
	V1	0	0	0	0	—
B	dACC	—	−3.55E−07	**−0.00259**	**−0.00387**	0
Congruent	rACC	−0.00005	—	0	0	0
	DLPFC	−0.00012	0	—	−0.00276	0
	VLPFC	−0.00042	0	0	—	0
	V1	0	0	0	0	—
C	Congruent	0	0	0	0	0.03473
	Incongruent	0	0	0	0	0.04094

Note: Figures represent strength of influence of the areas in the top row on those areas in the second column from the left. Matrix A: intrinsic connectivity; matrix B: connectivity modulated by stimuli; matrix C: activity modulated by stimuli. Parameters in bold font indicate increased influence of DLPFC and VLPFC on dACC in the incongruent condition compared to the congruent.

11.5.3   MULTIVARIATE AUTOREGRESSIVE MODELS

Goebel et al. (2003) and Harrison (2003) pioneered the use of multivariate autoregressive (MAR) models for the analysis of fMRI data. As previously described in the GCM section, the appeal of MAR is its ability to extract the temporal relations between signals. However, the analysis in GCM is carried out in a bivariate fashion (between the seed and each one of the other voxels), while for an accurate effective connectivity analysis, the signals of all ROIs should be included in a joint model. This multivariate joint estimation, which is present in the MAR model, is important in effective connectivity because the aim is to investigate and quantify the direct relationships between regions. The MAR model of order p (representing the number of previous time series) for K regions of interest $(A, B, \ldots, K)$ is given by

$$
\begin{cases}
A_t = a_{A1}A_{t-1} + \cdots + a_{Ap}A_{t-p} + \cdots + a_{K1}K_{t-1} + \cdots + a_{Kp}K_{t-p} + s_A u_{At} \\
B_t = b_{A1}A_{t-1} + \cdots + b_{Ap}A_{t-p} + \cdots + b_{K1}K_{t-1} + \cdots + b_{Kp}K_{t-p} + s_B u_{Bt} \\
\vdots \\
K_t = k_{A1}A_{t-1} + \cdots + k_{Ap}A_{t-p} + \cdots + k_{K1}K_{t-1} + \cdots + k_{Kp}K_{t-p} + s_K u_{Kt}.
\end{cases}
$$

In contrast to SEM, MAR are linear models which focus on the prediction of the BOLD signal at a certain region based on the past value (up to p time series previously) of the signal from all regions in the model. The MAR model is a specific type of multiple regression model, and can be estimated using the least-squares methods. The Granger causality test for the influence of region A on region K can be carried out by testing whether at least one coefficient $k_{A1}, \ldots, k_{Ap}$ is different from zero (F-test). Evaluation of the influence of any other region is accomplished in an analogous manner (e.g., influence of region K on region B is tested using coefficients $b_{K1}, \ldots, b_{Kp}$). It should be noted that this analysis infers the influence and relationships of regions based on the temporal precedence of their BOLD signals, taking into consideration the signals of all included ROIs. Since the BOLD signal is based on hemodynamic coupling, the results from this analysis must be interpreted with some caution, as discussed in Section 11.6.4.

Some alternative approaches to deal with the multivariate generalizations of VAR models based on conditional Granger causality were proposed by Zhou et al. (2009) and Liao et al. (2010). Conventional MAR models have some limitations, which are discussed in detail in Section 11.6. In brief, the assumption of stationarity of connectivity in data can be overcome with the dynamic MAR model (Sato et al., 2006, discussed in Section 11.6.3) and the complicating factors of cardiac and respiratory rhythms can be somewhat alleviated by frequency domain analysis such as partial directed coherence (Baccalá and Sameshima, 2001; Sato et al., 2009, discussed in Section 11.6.4).

As an illustrative example, we shall endeavor to infer Granger causality using the BOLD signals of the same ROIs considered in SEM analysis of the Stroop Test Data (Figure 11.7). Since MAR, in its original form, is limited to a single experimental condition, we will use MAR to model the whole session. Therefore, the BOLD variations we are modeling using MAR are those which occur in both congruent and

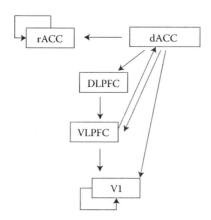

FIGURE 11.9 Estimated direction of information flow during the Stroop task (all conditions) using Granger causality analysis in MAR modeling. The model includes the rostral anterior cingulate cortex (rACC), dorsal anterior cingulate cortex (dACC), dorsolateral prefrontal cortex (DLPFC), ventrolateral prefrontal cortex (VLPFC), and primary visual cortex (V1).

incongruent conditions, and not those that distinguish between these two conditions. The R code for Granger causality tests based on MAR models is given by:

```
FIT=GrangerTEST(cbind(dACC,rACC,DLPFC,VLPFC,V1))
pvalue=1-pnorm(FIT)
```

Significant inter-region information flow ($p < 0.05$) is shown in Figure 11.9, and the p-values of each possible connection are described in Table 11.3. The major difference between this analysis and those using SEM and DCM is that the directions of causal influence were estimated from the data, in a full exploratory analysis.

The results of this exploratory analysis suggest the presence of a top-down modulation not observed using either SEM or DCM. This implies that previous values of

TABLE 11.3
p-Values of Granger Causality Tests Using the MAR Model on the Stroop Task Data Set

To\From	dACC	rACC	DLPFC	VLPFC	V1
dACC	**<0.001**	0.072	0.501	**<0.001**	0.68
rACC	**0.001**	0.094	0.956	0.439	0.17
DLPFC	**0.008**	0.536	0.765	0.196	0.682
VLPFC	**<0.001**	0.787	**0.04**	0.235	0.122
V1	**<0.001**	0.526	0.669	**<0.001**	**0.034**

Note: Figures represent the statistical significance of influence of the areas in the top row on those areas in the first column. The p-values in bold font indicate significant influence.

the BOLD signal at the frontal regions contain information about the current value at the visual cortex. In addition, it should be noted that the direction of links was estimated using signals with a sampling rate (TR) of 2 s. Therefore, connectivity between regions, which occurs on a smaller time scale, might not be observable in the data, or may seem "instantaneous." Any instantaneous influences will be present in the MAR model, but will not be highlighted by Granger causality, since direction of information flow would be impossible to infer. Consequently, correlation and SEM are more appropriate approaches for the quantification of these instantaneous influences. Thus, Granger causality analysis using MAR is a complementary approach to other effective connectivity analyses when exploration of temporal precedence relationships is required.

11.6 COMPARISON OF METHODS

11.6.1 DIRECTION OF INFORMATION FLOW

One of the most challenging issues for connectivity analysis methods applied to multivariate signals is the retrieval and estimation of the direction of information flow within a network. Approaches based on instantaneous linear dependence such as correlations provide only a symmetric measure of association between two regions and are consequently not capable of giving any insight about directionality in the data. SEM and DCM were developed with the aim of quantifying the strength of connectivity in a model with all connections in the network, and the direction of information flow in this network already specified. Although these methods can be useful for comparing different hypotheses in a confirmatory or model-driven analysis, they are not suitable in cases of exploratory or data-driven analysis. Extensions of these methods, which would enable more exploratory analysis, were proposed by Bullmore et al. (2000) for SEM and Friston et al. (2010) for DCM. However, the statistical and heuristic properties of these proposals, such as solution uniqueness, convergence, consistency, and evaluation of type I or type II, errors have not been established yet. Therefore, for exploratory analysis which allows assessment of the direction of information flow, a Granger causality-based approach is an appealing alternative. On the other hand, the direct application of conventional Granger causality test (as presented here) suffers from several limitations due to signal subsampling and regional/intersubject hemodynamic variability (Smith et al., 2011). Thus, caution is recommended when interpreting the results. A good review on the improvements and limitations of Granger causality application to fMRI data can be found in Stephan and Roebroeck (2012).

11.6.2 BIOLOGICAL VALIDITY

An important point raised by Friston et al. (2003) when they introduced DCM was that most proposals for connectivity analysis were actually making inferences about relationships between observed signals and not relationships between the neural activity of different regions. The interactions of most interest for researchers occur at the neuronal level, which is not directly observed in standard approaches. The origins of these methods in statistics and econometrics meant that neurophysiological information about the generation of the BOLD signal was being discarded. For this reason, DCM incorporates latent (unobserved) variables and the hemodynamic coupling process in its analyses. It uses these variables to estimate the realistic biophysical process that best describes the data. This is what differentiates DCM from all the other approaches, but the greater number of parameters and its nonlinear nature also makes for a much more complex analytical tool. Owing to this high complexity, it can lead to difficulties in interpretation, as there are questions asked about the possible solution uniqueness, convergence, and singularity problems within the method.

In contrast, the other approaches are only concerned with extracting relevant information about connectivity strength and direction, without attempting to fit a realistic biological model. This allows a more parsimonious analysis, with simpler and quicker results, but does not reflect the biology of the neural system as clearly. No one approach is perfect; instead, one must choose the best method for the specific data, problem, and hypothesis addressed.

11.6.3 COMPARISON OF CONNECTIVITY IN DIFFERENT EXPERIMENTAL CONDITIONS

In fMRI experiments, which involve the presentation of stimuli or the execution of certain tasks, the comparison of connectivity in different conditions (e.g. task vs. baseline) is often useful. There are a number of approaches that can be applied to investigate the effect of task condition. PPI is a good approach for functional connectivity analysis, and the intuitive extension for effective connectivity analysis is SEM. In addition, DCM directly models the effect of conditions in its extrinsic and modulatory parameters. However, Granger causality measurement based on conventional MAR would be unsuitable for this type of analysis, since it assumes stationarity of connectivity during the whole data set. As we are testing for task-based changes in connectivity and would expect some modulation of connections by experimental stimuli, this assumption is violated. The dynamic MAR model (Sato et al., 2006), which is based on wavelet expansion allowing a time-varying structure within MAR parameters, can be used to overcome this limitation.

11.6.4 LIMITATIONS OF FUNCTIONAL CONNECTIVITY ANALYSIS

As mentioned in the introduction, the increased use of fMRI is partially due to its high spatial resolution. However, it is limited by its low temporal resolution (sampling rate) as a result of its basis on the hemodynamic response. This may not be a problem if one is interested in locating voxels activated during certain tasks, but it could be important in connectivity analyses which investigate the temporal relationship between signals. In fact, the majority of techniques outlined in this chapter (SEM, PPI, and correlation analyses) are not based on time-series features, and are therefore not strongly affected by the sampling rate. Methods that are based on temporal precedence, such as Granger causality identification, may see a large reduction in their statistical power to detect connections at lower sampling rates. Nevertheless, in a study based on electrophysiological signals from a macaque and computational simulations, Deshpande et al. (2010) demonstrated that Granger causality analysis can detect temporal precedence of 100 ms, even in BOLD signals acquired at a sampling rate of 0.5 Hz (every 2 s). Schippers et al. (2011) studied the effects of variability within and between subjects on Granger causality analysis, and concluded that when a significant connectivity was identified between regions, the direction of influence within this connection was accurate in the majority of cases.

The basis of Granger causality is temporal precedence. However, there is a strong assumption fundamental to the application of this concept to fMRI data: the phase of hemodynamic response function must be the same for all voxels (in case of GCM) or within all ROIs (in case of effective connectivity analysis). The interactions which are being investigated occur between neurons, and the BOLD signal can be considered as a filtered version of this interaction and activity, convolved by the local hemodynamic response. Thus, if the phase of hemodynamic coupling is different between ROIs or voxels, temporal precedence is no longer a meaningful reflection of neuronal interaction. However, the low sampling rate of fMRI may help alleviate this problem. In most cases, the differences in hemodynamic delay are not expected to be greater than the typical sampling rate (TR) of 2 s. However, Roebroeck et al. (2009) have shown that the results of Granger causality analysis of fMRI signals can be greatly enhanced by deconvolving the BOLD signals at different regions by the respective estimated hemodynamic response function before further analysis. Actually, the reliability and limitations of Granger causality-based methods in fMRI are still controversial (Stephan and Roebroeck, 2012; Schippers et al., 2011; Smith et al., 2011).

A further complication with using the BOLD signal as an indicator of neural activity is that, as a hemodynamic process, it is strongly influenced by cardiac and respiratory rhythms. These physiological rhythms may drive common components in the BOLD signals measured at different ROIs, generating a spurious correlation in the time series. As mentioned in Section 11.4.1, there are methods which account for this complication in simple correlational analyses. If data on these rhythms have been acquired, the BOLD signal is frequently partialized by these confounder variables. Alternatively, analysis of the BOLD signal of voxels from a blood vessel can be used as a control. Multiple regression is also often applied by researchers needing to address confounds within their data. In addition, cardiac and respiratory oscillations have been reported to have a different frequency pattern and therefore a different frequency-related influence on resting-state correlations from the low frequencies (0.01–0.1 Hz) which are of interest (Cordes et al., 2001). It should therefore be possible to evaluate which frequency components are more related to the task, and which are related to physiological artifacts. As mentioned in Section 11.5.3, if you wish to perform analysis of temporal precedence on the data, frequency domain analysis of Granger causality, such as partial directed coherence, can be performed (Baccalá and Sameshima, 2001; Sato et al., 2009).

A crucial and still unsolved problem, common to all effective connectivity analysis, is the variation in number and location of nodes within a network. For example, omission of an important node in a network can lead to misspecification and very different connectivity results. It is worth mentioning that it is not always possible to include all the relevant regions within the model. This can be due to the numerical and overfitting problems caused by the large number of parameters to be estimated or the lack of an objective criterion to define the ROIs. Typically, effective connectivity analyses are performed using around three to six ROIs. Currently, all effective connectivity analyses assume that every region involved in the network has been specified and included in the model. Violations of this assumption may lead to biased estimates and spurious inferences. However, this does not mean that the methods are meaningless. The effects of variable omission may be subtle or large depending on the process investigated, the data in hand, and the node omitted. This limitation must be considered when interpreting results from effective connectivity analyses, and such interpretation must acknowledge the structure of the model that produced the results. These results may increase the body of evidence for a particular hypothesis, or suggest avenues for further research, but they cannot be seen as a proof or irrefutable demonstration of any theory.

Finally, recent studies recommend extreme caution when handling head micromovements. These studies have shown that this artifact can influence and bias the measures of functional connectivity, depending on the Euclidean distance between the regions of interest (Power et al., 2012; Satterthwaite et al., 2012). Thus, this limitation must be taken into account when comparing groups of subjects (e.g., healthy controls vs. patients).

11.7   CONCLUSION

In this chapter, we described some of the most widely used methods to infer brain connectivity from fMRI signals. These methods analyze the data and extract the information from different perspectives and they require different levels of assumptions. Exploratory and model-driven analyses can be carried out, depending on the questions to be answered and the amount of previous information. For this reason, all these methods should be seen as complementary and not as competitors. Future directions indicate the development of new methods based on multimodal integration, such as simultaneous EEG-fMRI or EEG-NIRS (near infra-red spectroscopy), in which joint information about metabolic and electrical processes can be obtained and analyzed.

REFERENCES

Biswal, B., F. Z. Yetkin, V. M. Haughton et al. 1995. Functional connectivity in the motor cortex of resting human brain using echo-planar MRI. *Magn. Reson. Med. 34*: 537–541.
Baccalá L. A. and K. Sameshima. 2001. Partial directed coherence: A new concept in neural structure determination. *Biol. Cybern. 84*: 463–474.
Beckmann C. F. and S. M. Smith. 2004. Probabilistic independent component analysis for functional magnetic resonance imaging. *IEEE Trans. Med. Imaging 23*: 137–152.
Buckner, R. L., J. R. Andrews-Hanna, and D. L. Schacter. 2008. The brain's default network: Anatomy, function and relevance to disease. *Ann. N.Y. Acad. Sci. 1124*: 1–38.
Bullmore, E., B. Horwitz, G. Honey et al. 2000. How good is good enough in path analysis of fMRI data? *Neuroimage 11*: 289–301.
Buxton, R. B., K. Uludağ, D. J. Dubowitz et al. 2004. Modeling the hemodynamic response to brain activation. *Neuroimage 23*: S220–S233. Review.

Comon, P. 1994. Independent component analysis, A new concept? *Signal Process. 36*: 287–314.

Cordes, D., V. M. Haughton, K. Arfkanis et al. 2001. Frequencies contributing to functional connectivity in the cerebral cortex in "resting-state" data. *Am. J. Neuroradiol. 22*: 1326–1333.

Deshpande, G., K. Sathian, and X. Hu. 2010. Effect of hemodynamic variability on Granger causality analysis of fMRI. *Neuroimage 52*: 884–896.

Friston, K. J. 1994. Functional and effective connectivity in neuroimaging: A synthesis. *Hum. Brain Mapp. 2*: 56–78.

Friston, K. J., C. Buechel, G. R. Fink et al. 1997. Psychophysiological and modulatory interactions in neuroimaging. *Neuroimage 6*: 218–229.

Friston, K. J., L. Harrison, and W. Penny. 2003. Dynamic causal modelling. *Neuroimage 19*: 1273–1302.

Friston, K. J., B. Li, J. Daunizeau et al. 2010. Network discovery with DCM. *Neuroimage 56*: 1202–1221.

Goebel, R., A. Roebroeck, D. S. Kim et al. 2003. Investigating directed cortical interactions in time-resolved fMRI data using vector autoregressive modeling and Granger causality mapping. *Magn. Reson. Imaging 21*: 1251–1261.

Granger, C. W. J. 1969. Investigating causal relations by econometric models and cross-spectral methods. *Econometrica 37*: 424–438.

Harrison, L., W. D. Penny, and K. Friston. 2003. Multivariate autoregressive modeling of fMRI time series. *Neuroimage 19*: 1477–1491.

Hyvörinen, A. 1999. Fast and robust fixed-point algorithms for independent component analysis. *IEEE Trans. Neural Networks 10*: 626–634.

Jezzard, P., P. M. Matthews, and S. M. Smith. 2003. *Functional MRI: An Introduction to Methods*, 1st ed. Oxford: Oxford University Press.

Liao, W., D. Mantini, Z. Zhang et al. 2010. Evaluating the effective connectivity of resting state networks using conditional Granger causality. *Biol. Cybern. 102*: 57–69.

Logothetis, N. K., J. Pauls, M. Augath et al. 2001. Neurophysiological investigation of the basis of the fMRI signal. *Nature 412*: 150–157.

McIntosh, A. R. 1998. Understanding neural interactions in learning and memory using functional neuroimaging. *Ann. N.Y. Acad. Sci. 855*: 556–571.

McKeown, M. J., S. Makeig, G. G. Brown et al. 1998. Analysis of fMRI data by blind separation into independent spatial components. *Hum. Brain Mapp. 6*: 160–88.

Ogawa, S., T. M. Lee, A. R. Kay et al. 1990. Brain magnetic resonance imaging with contrast dependent on blood oxygenation. *Proc. Natl. Acad. Sci. USA 87*: 9868–9872.

Penny, W. D., K. J. Friston, J. T. Ashburner et al. 2006. *Statistical Parametric Mapping: The Analysis of Functional Brain Images*. London: Academic Press.

Peterson, B. S., P. Skudlarski, J. C. Gatenby et al. 1999. An fMRI study of Stroop word-color interference: Evidence for cingulate subregions subserving multiple distributed attentional systems. *Biol. Psychiatry 45*: 1237–1258.

Power, J. D., K. A. Barnes, A. Z. Snyder et al. 2012. Spurious but systematic correlations in functional connectivity MRI networks arise from subject motion. *Neuroimage 59*: 2142–2154.

Roebroeck, A., E. Formisano, and R. Goebel. 2011. The identification of interacting networks in the brain using fMRI: Model selection, causality and deconvolution. *Neuroimage 58*: 296–302.

Rowe, J. B. 2010. Connectivity analysis is essential to understand neurological disorders. *Front. Syst. Neurosci. 4*: 144.

Sato, J. R., E. Amaro Jr, D. Y. Takahashi et al. 2006. A method to produce evolving functional connectivity maps during the course of an fMRI experiment using wavelet-based time-varying Granger causality. *NeuroImage 31*: 187–196.

Sato, J. R., D. Y Takahashi, S. M. Arcuri et al. 2009. Frequency domain connectivity identification: An application of partial directed coherence in fMRI. *Hum. Brain Mapp. 30*: 452–461.

Satterthwaite, T. D., D. H. Wolf, J. Loughead et al. 2012. Impact of in-scanner head motion on multiple measures of functional connectivity: Relevance for studies of neurodevelopment in youth. *NeuroImage 60*: 623–632.

Schippers, M. B., R. Renken, and C. Keysers. 2011. The effect of intra-and inter-subject variability of hemodynamic responses on group level Granger causality analyses. *NeuroImage 57*: 22–36.

Schlösser, R. G., G. Wagner, K. Koch et al. 2008. Fronto-cingulate effective connectivity in major depression: A study with fMRI and dynamic causal modeling. *NeuroImage 43*: 645–655.

Simon, H. 1953. Causal ordering and identifiability. In *Studies in Econometric Method*, eds. W. C. Hood, and T. C. Koopmans, 49–74. New York: Wiley.

Smith, S. M., K. L. Miller, G. Salimi-Khorshidi et al. 2011. Network modelling methods for fMRI. *NeuroImage 54*: 875–891.

Stephan, K. E. and A. Roebroeck. 2012. A short history of causal modeling of fMRI data. *NeuroImage 62*: 856–863.

van den Heuvel, M. P. and H. E. Hulshoff Pol. 2010. Exploring the brain network: A review on resting-state fMRI functional connectivity. *Eur. Neuropsychopharmacol. 20*: 519–534.

Wright, S. S. 1921. Correlation and causation. *J. Agric. Res. 20*: 557–585.

Zhou, Z., Y. Chen, M. Ding et al. 2009. Analyzing brain networks with PCA and conditional Granger causality. *Hum. Brain Mapp. 30*: 2197–2206.

Zoccatelli, G., A. Beltramello, F. Alessandrini et al. 2010. Word and position interference in Stroop tasks: A behavioral and fMRI study. *Exp. Brain Res. 207*: 139–147.

12 Assessing Causal Interactions among Cardiovascular Variability Series through a Time-Domain Granger Causality Approach

Alberto Porta, Anielle C. M. Takahashi, Aparecida M. Catai, and Nicola Montano

CONTENTS

12.1   INTRODUCTION

Cardiovascular variables exhibit spontaneous, nonrandom, beat-by-beat fluctuations about their mean value and these variations are referred to as cardiovascular variability (Cohen and Taylor, 2002). Since the magnitude of these fluctuations depends on the state of the autonomic nervous system (Akselrod et al., 1981; Malliani et al., 1991), spontaneous cardiovascular variability has been basically exploited to noninvasively infer it. For example, the amount of the fluctuations of the heart period (HP) in the high frequency (HF, around the respiratory rate) band is considered as an index of vagal modulation directed to the heart, while the magnitude of the fluctuations of systolic arterial pressure (SAP) in the low-frequency (LF, from 0.04 to 0.15 Hz) band is generally employed as an index of sympathetic modulation directed toward the vessels (Akselrod et al., 1981; Cooke et al., 1999; Pagani et al., 1986). Since the evaluation of the autonomic control is mainly based on the assessment of the magnitude of HP and SAP fluctuations in the LF and HF bands, spectral analysis is the most widely utilized tool in cardiovascular variability studies.

This traditional approach to the noninvasive assessment of the autonomic control was found very useful in patho-physiological studies (Kleiger et al., 1987; Task Force, 1996, 2001; La Rovere et al., 2003; Guzzetti et al., 2005), but it leaves completely unsolved the puzzle of the causal interactions among cardiovascular variables. The issue of causality stimulates fierce debates in the cardiovascular field (Eckberg and Karemaker, 2009). For example, the presence of highly correlated HP and SAP oscillations at the respiratory rate has been contemporaneously interpreted as an indication that HP fluctuations, induced by central respiratory gating of vagal motoneuron responsiveness, contribute to SAP changes (Eckberg, 2003) and as a suggestion of the reverse causal relation (i.e., SAP fluctuations drive HP changes via a baroreflex) (De Boer et al., 1987). Solving the puzzle of the causal interactions is not only important to better understand the mechanisms responsible for linking cardiovascular variables but also has practical consequences on the estimation of cardiovascular parameters such as baroreflex sensitivity (Smyth et al., 1969; La Rovere et al., 2001). For example, when baroreflex sensitivity is assessed from spontaneous HP and SAP variabilities (Laude et al., 2004), its estimate is reliable only if SAP fluctuations at the respiratory rate contribute to HP changes (Porta et al., 2000a).

Another relevant limitation of the traditional approach based on spectral analysis lies in its inability to test hypotheses that are implicitly assumed in interpreting and modeling cardiovascular variability interactions. For example, it is commonly believed that respiration (R) is an exogenous source affecting cardiovascular variables (Kitney et al., 1985; De Boer et al., 1987; Baselli et al., 1994; Xiao et al., 2005) without being affected. However, it is well known that the stimulation of central and peripheral chemoreceptors causes sudden HP variations driven by sympathetic circuits that might precede hyperventilation, thus suggesting a causal link from HP to R (Kara et al., 2003). Even in case of an unlikely activation of chemoreflex, it was observed that an HP decrease might precede inspiration when monitored via two-belt chest–abdomen inductance plethysmography (Yana et al., 1993; Perrott and Cohen, 1996), especially when the breathing rate is slow (<0.15 Hz) (Saul et al., 1989). The simplistic hypothesis of unidirectional interaction from R to HP and SAP is never rigorously tested, thus preventing the judgment of the consistency of this assumption over the short-term recordings of cardiovascular variables.

The aim of this study is to assess causal relations among cardiovascular variabilities during an experimental protocol capable of gradually modifying the state of the autonomic nervous system. We tested the hypothesis that an autonomic nervous system plays a role in picking and/or turning off specific causality schemes and that cardiovascular control follows the strategy of selecting causality patterns to guarantee a rapid and flexible reaction to mutable internal and external conditions. Causality was assessed according to the Granger causality approach (Granger, 1969) stating that y_k Granger-causes y_m if y_k carries a unique information about the future evolution of y_m that cannot be derived from any other series forming the universe of knowledge about the system under study. Granger's causality was assessed using an appropriate F-test (Söderström and Stoica, 1988) by comparing the predictability of a series (here HP, SAP, or R) derived according to the most complete information set describing the system functioning (i.e., the recordings of HP, SAP, and R series) with that derived after excluding the series supposed to be the cause from the information set.

12.2   METHODS

12.2.1   Assessing Granger Causality in the Time Domain

Given the series $Y_1 = \{Y_1(n), n = 1, \ldots, N\}$, $Y_2 = \{Y_2(n), n = 1, \ldots, N\}, \ldots, Y_M = \{Y_M(n), n = 1, \ldots, N\}$ where n is the progressive counter and N is the series length, first the means are removed and, then, the resulting values are divided by the standard deviation, thus obtaining the $y_1, y_2, \ldots, y_M$ series with a zero mean and unit variance. The set of M signals, $\Omega_y = \{y_1, \ldots, y_{m-1}, y_m, y_{m+1}, \ldots, y_M\}$, represents the universe of our knowledge about the system under examination on the hypothesis that the interactions of the M signals can be fully described in Ω_y (i.e., completeness of Ω_y) and each of the signals carries a certain amount of information about the system that cannot be completely derived from any other signal in Ω_y (i.e., no redundancy in Ω_y). The dynamics of y_k with $1 \leq k \leq M$ can be described (Lütkepohl, 2005) as the sum of a linear combination of past values of y_k, a linear mixture of present and past values of all the remaining signals in Ω_y, and a random input w_k describing the portion of y_k that cannot be predicted in Ω_y. Thus,

$$y_k(n) = A_{kk}(z)\bigg|_{\Omega_y} y_k(n) + \sum_{l=1, l \neq k}^{M} A_{kl}(z)\bigg|_{\Omega_y} y_l(n) + w_k(n)\bigg|_{\Omega_y}, \qquad (12.1)$$

where

$$A_{kk}(z)\bigg|_{\Omega_y} = \sum_{j=1}^{q} a_{kk}(j)\bigg|_{\Omega_y} z^{-j} \qquad (12.2)$$

describes the dependence of y_k on its own past values in Ω_y ($1 \le k \le M$) and

$$A_{kl}(z)\Big|_{\Omega_y} = \sum_{j=\tau_{kl}}^{q} a_{kl}(j)\Big|_{\Omega_y} z^{-j} \qquad (12.3)$$

describes the dependence of y_k on the present and past values of y_l in Ω_y ($l \ne k$ and $1 \le k, l \le M$), where z^{-1} is the backward shift operator in the z-domain (i.e., $z^{-1}y_k(n) = y_k(n-1)$), τ_{kl} is the delay of influence of y_l on y_k, and q is the model order. While Equation 12.2 suggests that the immediate effects of y_k on itself are not allowed (i.e., $a_{kk}(0) = 0$), Equation 12.3 suggests that the instantaneous effects of y_l on y_k might be allowed by setting $\tau_{kl} = 0$. The condition $a_{kk}(0) = 0$ avoids the impractical immediate action of $y_k(n)$ over itself, while the choice of τ_{kl} depends on physiological considerations about the relation of y_l on y_k. If $\tau_{kl} = 0$ is set, special attention must be paid to avoid the formation of loops without delays. If both $\tau_{kl} \ne 0$ and $\tau_{lk} \ne 0$, the immediate effects are disregarded.

By defining $\Omega_y - \{y_m\} = \{y_1, \ldots, y_{m-1}, y_{m+1}, \ldots, y_M\}$ as the universe Ω_y after excluding y_m, y_k with $1 \le k \le M, k \ne m$ can be written as

$$y_k(n) = A_{kk}(z)\Big|_{\Omega_y - \{y_m\}} y_k(n) + \sum_{l=1, l \ne k, m}^{M} A_{kl}(z)\Big|_{\Omega_y - \{y_m\}}$$

$$\times\, y_l(n) + w_k(n)\Big|_{\Omega_y - \{y_m\}} \qquad (12.4)$$

representing the description of y_k given by Equation 12.1 when y_m is not accounted for. The coefficient of A_{kl} with $1 \le k, l \le M$ can be identified according to some optimization criterion applied to the single autoregressive process with exogenous inputs, y_k (e.g., the minimization of the variance of w_k) or to the multivariate autoregressive process $y = |y_1 \cdots y_M|^T$ (e.g., the determinant of the variance matrix of $w = |w_1 \cdots w_M|^T$) (Söderström and Stoica, 1988). The prediction of y_k in Ω_y is

$$\hat{y}_k(n)\Big|_{\Omega_y} = \hat{A}_{kk}(z)\Big|_{\Omega_y} y_k(n) + \sum_{l=1, l \ne k}^{M} \hat{A}_{kl}(z)\Big|_{\Omega_y} y_l(n) \qquad (12.5)$$

whereas that in $\Omega_y - \{y_m\}$ is

$$\hat{y}_k(n)\Big|_{\Omega_y - \{y_m\}} = \hat{A}_{kk}(z)\Big|_{\Omega_y - \{y_m\}} y_k(n) + \sum_{l=1, l \ne k, m}^{M} \hat{A}_{kl}(z)\Big|_{\Omega_y - \{y_m\}} y_l(n). \qquad (12.6)$$

Defining the prediction error, $e(n)$, as the difference between $y_k(n)$ and its best prediction, $\hat{y}_k(n)$, the mean square prediction error over the N samples can be assessed in Ω_y and in $\Omega_y - \{y_m\}$ and indicated as $\hat{\lambda}_k^2\Big|_{\Omega_y}$ and $\hat{\lambda}_k^2\Big|_{\Omega_y - \{y_m\}}$, respectively.

According to the Granger causality approach, y_m Granger-causes y_k in Ω_y, in the following indicated as $y_m \rightarrow y_k$, if $\hat{\lambda}_k^2 \Big|_{\Omega_y - \{y_m\}}$ is significantly larger than $\hat{\lambda}_k^2 \Big|_{\Omega_y}$ (i.e., the exclusion of y_m from Ω_y worsens $\hat{y}_k$) (Granger, 1969).

12.2.2 F-TEST

Among the possible tests to check whether a model over Ω_y fits better y_k than that over $\Omega_y - \{y_m\}$ (Eberts and Steece, 1984), the traditional F-test is largely utilized given its simplicity (Söderström and Stoica, 1988). Indeed, F-test is simply based on a comparison between $\hat{\lambda}_k^2 \Big|_{\Omega_y}$ and $\hat{\lambda}_k^2 \Big|_{\Omega_y - \{y_m\}}$ according to the ratio

$$F_{km} = \frac{\left(\hat{\lambda}_k^2 \Big|_{\Omega_y - \{y_m\}} - \hat{\lambda}_k^2 \Big|_{\Omega_y} \right)}{\hat{\lambda}_k^2 \Big|_{\Omega_y}} \cdot \frac{\left[N - q - (M-1)(q+1) + \sum_{l=1, l \neq k}^{M} \tau_{kl} \right]}{(q + 1 - \tau_{km})}, \quad (12.7)$$

where τ_{km} represents the delay of influence of y_m on y_k.

If F_{km} is larger than the critical value of the F distribution derived from a given significance level (the degrees of freedom of the numerator and denominator are equal to $q + 1 - \tau_{km}$ and $N - q - (M-1)(q+1) + \sum_{l=1, l\neq k}^{M} \tau_{kl}$, respectively), the null hypothesis that y_m does not Granger-cause y_k is rejected and the alternative hypothesis of $y_m \rightarrow y_k$ is accepted (unidirectional causality from y_m to y_k). Reversing the role of y_m and y_k allows the assessment of $y_k \rightarrow y_m$. If both $y_m \rightarrow y_k$ and $y_k \rightarrow y_m$ are contemporaneously found, a closed loop relation (bidirectional causality) can be argued (i.e., $y_m \leftrightarrow y_k$). If the null hypothesis that y_m does not Granger-cause y_k and vice versa cannot be rejected, y_m and y_k are uncoupled. The significance level was set at $\alpha = 0.01$, thus rejecting the null hypothesis with a probability of type I error, $p < 0.01$.

12.2.3 DIRECTIONALITY INDEX

The assessment of the dominant causality can be based on a direct comparison between directionality indexes assessed over opposite causal direction such as F_{km} and F_{mk}. Accordingly, the directionality index (DI) (Rosenblum and Pikovsky, 2001; Paluš and Stefanovska, 2003) is defined as

$$DI_{km} = F_{km} - F_{mk}. \quad (12.8)$$

$DI_{km} > 0$ indicates that the causal direction from y_m to y_k is prevalent over the reverse one, while $DI_{km} < 0$ points out the opposite situation. DI_{km} is exclusively capable of identifying the dominant causality: indeed, $DI_{km} > 0$ or $DI_{km} < 0$ does not exclude bidirectional interactions. In addition, DI_{km} close to 0 might indicate: (i) a full uncoupling between y_k and y_m; (ii) closed loop interactions between y_k and y_m

with none of the causal directions taking real preeminence; and (iii) synchronization between y_k and y_m.

12.3 EXPERIMENTAL PREPARATION AND DATA ANALYSIS

12.3.1 DATA RECORDINGS

We studied 19 healthy humans (aged 21–48, median = 30; 11 females and 8 males). ECG (lead II), continuous arterial pressure (Finometer MIDI, Finapres Medical Systems, The Netherlands), and respiratory movements via a thoracic belt (Marazza, Monza, Italy) were recorded. Signals were sampled at 300 Hz. All the experimental sessions included three periods of acquisition in the order and duration as follows: (1) 7 min in a supine position at rest (REST); (2) 10 min during a passive head-up tilt; and (3) 8 min of recovery. The inclination of the tilt table, expressed as degrees, was randomly chosen within the set $\{15, 30, 45, 60, 75, 90\}$ (T15, T30, T45, T60, T75, T90). Each subject completed the sequence of tilt table angles. After 10 min of adaptation, the three periods were continuously recorded. The gradual head-up tilt protocol induces an increase of sympathetic activity and modulation (i.e., the amplitude of sympathetic activity changes about its mean value) proportional to the magnitude of the gravitational stimulus (Montano et al., 1994; Cooke et al., 1999; Furlan et al., 2000; Porta et al., 2007). The protocol adhered to the principles of the Declaration of Helsinki. The human research and ethical review boards of the "L. Sacco" Hospital and of the Department of Biomedical Sciences for Health approved the protocol.

12.3.2 SERIES EXTRACTION

Analyses of sessions were performed after about 2 min from their start. The QRS apex was located using parabolic interpolation and HP was approximated as the time interval between two consecutive QRS peaks. The location of the QRS peak was fixed according to a parabolic interpolation to improve time resolution of the fiducial point above and beyond the sampling period. The maximum of arterial pressure inside the i-th HP was taken as the i-th SAP. The respiratory movement signal was down-sampled once per cardiac beat at the occurrence of the first QRS peak delimiting the i-th HP, thus obtaining the i-th R measure. After extracting the series HP = $\{HP(i), i = 1, \ldots, N\}$, SAP = $\{SAP(i), i = 1, \ldots, N\}$, and $R = \{R(i), i = 1, \ldots, N\}$, where i is the progressive cardiac beat index and N is the total cardiac beat number, sequences of 256 consecutive measures were randomly selected, thus focusing on the short-term cardiovascular control mechanisms (Task Force, 1996). HP, SAP, and R were expressed in ms, mmHg, and arbitrary units (a.u.), respectively.

12.3.3 IDENTIFICATION OF THE MODEL PARAMETERS FROM CARDIOVASCULAR VARIABILITIES

Given that $y_1 = hp$, $y_2 = sap$, and $y_3 = r$, the coefficients of the polynomials of $A_{kl}(z)$ with $1 \leq k, l \leq 3$ were identified both in $\Omega_y = \{y_1, y_2, y_3\}$ and in $\Omega_y - \{y_k\}$

with $k = 1, 2, 3$ directly from the cardiovascular series using the traditional least-squares approach and the Cholesky decomposition method (Söderström and Stoica, 1988; Kay, 1989; Baselli et al., 1997). The delays were set to allow the description of the fast vagal reflex, within the same cardiac beat capable of modifying HP in response to changes of SAP and R (i.e., $\tau_{12} = \tau_{13} = 0$) (Eckberg, 1976; Porta et al., 2000a), of the rapid effect of R on SAP due to the immediate transfer of an alteration of the intrathoracic pressure on SAP value (i.e., $\tau_{23} = 0$) (Cohen and Taylor, 2002) and of the one-beat delayed effect of HP on SAP due to the measurement conventions preventing that the i-th HP could modify the i-th SAP (i.e., $\tau_{21} = 1$) (Baselli et al., 1994). According to Saul et al. (1989) and Saul et al. (1991), actions of HP and SAP on R were slower (i.e., they cannot occur in the same beat), thus leading to $\tau_{31} = 1$ and $\tau_{32} = 1$. The model order, q, was optimized in the 4–16 range according to the Akaike figure of merit for multivariate processes (Akaike, 1974). The coefficients of $A_{kl}(z)$ were obtained by minimizing the determinant of the variance matrix of $w = |w_1 \ w_2 \ w_3|^T$. Whiteness of the residuals, w_k with $1 \le k \le 3$, and their uncorrelation, even at zero lag, were tested. The optimal model order chosen in Ω_y was maintained even in $\Omega_y - \{y_k\}$ with $k = 1, 2, 3$ and the coefficients of the polynomials were estimated again.

12.3.4 STATISTICAL ANALYSIS

Linear regression analysis of the HP and SAP means, HP and SAP variances, F-values (i.e., F_{12}, F_{21}, F_{13}, F_{31}, F_{23}, and F_{32}), and directionality index (i.e., D_{21}, D_{31}, and D_{32}) on tilt angles was carried out. Pearson's product moment correlation coefficient, r, was calculated. A $p < 0.01$ was considered significant.

12.4 RESULTS

Figure 12.1 shows the HP, SAP, and R series recorded at REST (Figure 12.1a,b,c) and during T90 (Figure 12.1d,e,f). In Figure 12.1, the HP mean, μ_{HP}, was remarkably lower during T90 (Figure 12.1d) than at REST (Figure 12.1a), while the SAP variance, σ_{SAP}^2, was larger (Figure 12.1e vs. Figure 12.1b). Over the entire set of data, HP mean and variance, μ_{HP} and σ_{HP}^2, were linearly related to tilt table angles. Correlation of μ_{HP} and σ_{HP}^2 on the tilt table inclination was negative ($r = -0.57$ and $r = -0.31$, respectively) and significant (i.e., $p = 5.69 \times 10^{-13}$ and $p = 3.39 \times 10^{-4}$, respectively). SAP mean and variance, μ_{SAP} and σ_{SAP}^2, were significantly linearly related to the tilt table angle as well, but their correlation with tilt table angles was positive and weaker ($r = 0.24$, $p = 6.46 \times 10^{-3}$ and $r = 0.28$, $p = 1.28 \times 10^{-3}$, respectively).

Results of causality analysis between y_1 and y_2 are shown in Figure 12.2. While the F-value assessing causality from y_2 to y_1, F_{12}, progressively increased as a function of the tilt table angle (Figure 12.2a, $r = 0.29$ with $p = 6.94 \times 10^{-4}$), the F-value evaluating causality from y_1 to y_2, F_{21}, gradually decreased (Figure 12.2b, $r = -0.25$ with $p = 3.37 \times 10^{-3}$), thus suggesting that the predictability improvement of y_1 due to the introduction of y_2 gradually increased, while that of y_2 due to the introduction of y_1 progressively decreased. DI_{21} gradually decreased from positive values to negative

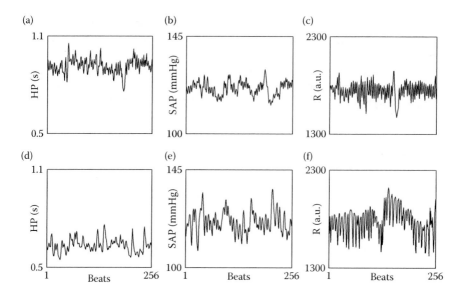

FIGURE 12.1 Series of HP (a,d), SAP (b,e), and R (c,f) recorded from the same subject at REST (a,b,c) and during T90 (d,e,f).

ones (Figure 12.2c, $r = -0.39$ with $p = 4.03 \times 10^{-6}$), thus indicating the gradual transition from a dominant causality $y_1 \rightarrow y_2$ to the reverse causal direction along the baroreflex pathway (i.e., $y_2 \rightarrow y_1$).

Results of causality analysis between y_3 and y_1 are shown in Figure 12.3. The F-value assessing causality from y_3 to y_1, F_{13} (Figure 12.3a) and from y_1 to y_3, F_{31} (Figure 12.3b) progressively decreased as a function of the tilt table inclination ($r = -0.26$ with $p = 2.44 \times 10^{-3}$ and $r = -0.39$ with $p = 3.08 \times 10^{-6}$, respectively), thus suggesting that the predictability improvement of y_1 due to the introduction of y_3 and that of y_3 due to the introduction of y_1 gradually decreased. DI_{31} fluctuated about 0 in all the experimental conditions (Figure 12.3c), thus indicating the absence

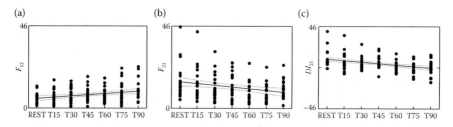

FIGURE 12.2 Individual values (solid circles) of (a) F_{12}, (b) F_{21}, and (c) D_{21} as a function of the tilt table inclination. The linear regressions (solid line) calculated over all the values and their 95% confidence interval (dotted lines) are plotted as well when the slope of the regression line was found to be significantly different from 0 with $p < 0.01$. F_{12}, F_{21}, and D_{21} were linearly correlated with the tilt table inclination.

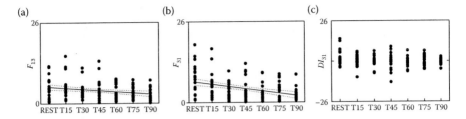

FIGURE 12.3 Individual values (solid circles) of (a) F_{13}, (b) F_{31}, and (c) D_{31} as a function of the tilt table inclination. The linear regressions (solid line) calculated over all the values and their 95% confidence interval (dotted lines) are plotted as well when the slope of the regression line was found to be significantly different from 0 with $p < 0.01$. Only F_{13} and F_{31} were linearly correlated with the tilt table inclination.

of a dominant causal relation in the interactions between y_1 and y_3. DI_{31} was not linearly related to the tilt table angle.

Results of causality analysis between y_3 and y_2 are shown in Figure 12.4. The F-value assessing causality from y_3 to y_2, F_{23} (Figure 12.4a) was significantly larger than that assessing causality from y_2 to y_3, F_{32} (Figure 12.4b) in all the experimental conditions. As a result, DI_{32} was always negative (Figure 12.4c), thus indicating the presence of a dominant causal relation from y_3 to y_2. F_{23}, F_{32}, and DI_{32} were not linearly related to the tilt table angle.

Figure 12.5 shows the rates of detection of causality patterns as a function of the experimental condition. Closed-loop relation between y_1 and y_2, $y_1 \leftrightarrow y_2$, was found in a large percentage of subjects at REST (i.e., 84%) and this fraction remained stable with the tilt table angle (Figure 12.5a, solid bars). The percentage of subjects with unidirectional causality (i.e., $y_2 \rightarrow y_1$ or $y_1 \rightarrow y_2$) or with uncoupling between y_1 and y_2 was negligible (Figure 12.5a) in all experimental conditions. At REST, the percentage of subjects with closed relation between y_1 and y_3, $y_1 \leftrightarrow y_3$ (Figure 12.5b, solid bars) was large (i.e., 63%), while y_1 and y_3 were found to be uncoupled (Figure 12.5b, open bar) in a small percentage of subjects (i.e., 16%). The percentage

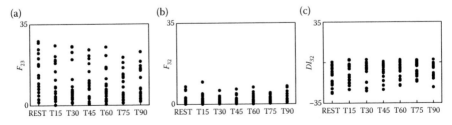

FIGURE 12.4 Individual values (solid circles) of (a) F_{23}, (b) F_{32}, and (c) D_{32} as a function of the tilt table inclination. The linear regressions (solid line) calculated over all the values and their 95% confidence interval (dotted lines) are plotted as well when the slope of the regression line was found to be significantly different from 0 with $p < 0.01$. None of the parameters was linearly correlated with the tilt table inclination.

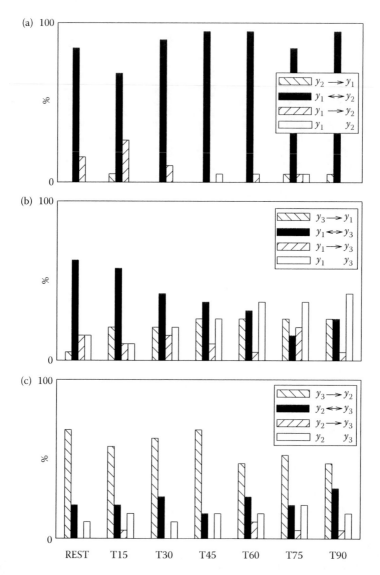

FIGURE 12.5 Percentage of (a) $y_1 - y_2$, (b) $y_1 - y_3$, and (c) $y_2 - y_3$ causal interactions during a graded head-up tilt protocol with $y_1 = $ hp, $y_2 = $ sap, and $y_3 = r$. The percentages of $y_m \rightarrow y_k$ (backslash-pattern bar), $y_k \leftrightarrow y_m$ (solid bar), $y_k \rightarrow y_m$ (slash-pattern bar), and uncoupled y_k and y_m (open bar) are shown in (a) with $m = 2$ and $k = 1$, in (b) with $m = 3$ and $k = 1$, and in (c) with $m = 3$ and $k = 2$.

of subjects with $y_1 \leftrightarrow y_3$ progressively decreased as a function of the tilt table inclination, while the uncoupling between y_1 and y_3 was found more frequently at the high tilt table angles. The percentage of subjects exhibiting unidirectional causality (i.e., $y_3 \rightarrow y_1$ or $y_1 \rightarrow y_3$) was not affected by the tilt table angle. At REST, the percentage of subjects with unidirectional causality from y_3 to y_2 was high (i.e., 68%)

(Figure 12.5c), bidirectional causality between y_2 to y_3 (i.e., $y_2 \leftrightarrow y_3$) and uncoupling between y_2 to y_3 were found in a small percentage of subjects (i.e., 21% and 10%, respectively), and unidirectional causality from y_2 to y_3 (i.e., $y_2 \rightarrow y_3$) was never detected. The percentage of all patterns defining the causal interactions between y_2 and y_3 remained stable with the tilt table inclination (Figure 12.5c).

12.5   DISCUSSION

The main findings of this study can be summarized as follows: (i) Granger causality approach in the time domain detected the gradual shift from a dominant causal relation from HP to SAP to the reverse causal link as a function of the sympathetic activation induced by a head-up tilt; (ii) this result was found in the presence of an extended information set including R with HP and SAP; (iii) R drove HP, but the reverse situation was also found (i.e., HP acted upon R); (iv) closed-loop relations between HP and R gradually weakened with sympathetic activation, thus leading to a dominant condition of HP-R uncoupling at the highest tilt table angles; (v) unidirectional interaction from R to SAP was more frequently found than the reverse causal link (i.e., from SAP to R); and (vi) the causal relation from R to SAP was independent of the level of autonomic activity and/or modulation.

12.5.1   CARDIOVASCULAR VARIABILITY STUDIES CAN BENEFIT FROM THE APPLICATION OF CAUSALITY APPROACHES

Cardiovascular variability series exhibit oscillations in specific frequency bands and the amplitude of these oscillations are commonly assessed using spectral analysis to understand cardiovascular physiology, typify populations, and stratify patients (Akselrod et al., 1981; Malliani et al., 1991; Task Force, 1996; La Rovere et al., 2003; Guzzetti et al., 2005). The richness of the dynamical patterns present in cardiovascular variability triggered the application of entropy-based complexity analysis with the aim of linking the amount of information carried by a signal with a variety of different mechanisms and circuits responsible for cardiovascular control (Pincus et al., 1993; Porta et al., 2000b; Richman and Moorman, 2000). Any monovariate tool, such as spectral or complexity analyses, treats a multivariate recording as a sequence of independent single series, thus being particularly helpful to describe dynamical features such as rhythmical patterns or information content, but being completely useless to describe interactions. Cross-spectral analysis is largely utilized in cardiovascular variability studies to overcome the limitation of monovariate methods due to its ability to derive the gain, phase, and strength of the relation between a pair of signals (De Boer et al., 1985; Saul et al., 1991; Cooke et al., 1999; Badra et al., 2001). Unfortunately, cross-spectral analysis, like any other bivariate approach, fails when more than two series are contemporaneously recorded, thus leaving the description of complexity of the joint interactions in multivariate recordings largely unaccounted. Multivariate model-based approaches based on system identification procedures (Xiao et al., 2005; Porta et al., 2009) provide a joint description of the multivariate interactions according to the structure of the selected model. Although multivariate model-based approaches provided the basis of causality analysis (Granger, 1969), applications of system identification modeling techniques in the field of cardiovascular variabilities did not face directly this issue of causality. Indeed, they were traditionally focused on the assessment of transfer functions, on the decomposition of the series into partial processes, on the classification of the rhythms present in the series, and on the estimation of the coupling strength (Baselli et al., 1994, 1997; Patton et al., 1996; Perrott and Cohen, 1996; Kim and Khoo, 1997; Mullen et al., 1997; Mukkamala et al., 2003; Porta et al., 2009). As a practical consequence, the causal relations in cardiovascular variability studies are still largely unexplored and tools typifying causality are little exploited. Some original applications can be found in Faes et al. (2001), Nollo et al. (2002), Porta et al. (2002), Nollo et al. (2005), Riedl et al. (2010), and Porta et al. (2011b). Thus, it is not surprising that, despite the large amount of tools exploring causality, a typical causality issue such as whether respiratory sinus arrhythmia contributes to SAP changes at the respiratory rate or vice versa (Eckberg and Karemaker, 2009) was faced by estimating phases between HP and SAP series. Unfortunately, phases are of little help with causality due to the presence of positive and negative phase multiples leading to ambiguities in the estimation of causality. These ambiguities are difficult to solve even when, after transformation of phases into delays or advancements, they are compared with the latency of the physiological reflex or mechanism (Porta et al., 2011b). Indeed, an accurate estimate of the latency is infrequent, thus preventing a precise exclusion of unlikely phase multiples and occasionally even the decision between delays or advancements. In addition, since phase estimation between two signals might be biased by the action of a third signal affecting both, decomposition into partial processes is necessary (Baselli et al., 1997; Badra et al., 2001; Porta et al., 2011a) to remove the joint influences and limiting phase biases.

12.5.2   DEFINITION OF THE INFORMATION SET FOR CAUSALITY ANALYSIS IN CARDIOVASCULAR VARIABILITY STUDIES

In this study, we utilized a Granger causality approach to explore causal relations. The Granger causality approach exploits an operational definition of causality that strictly depends on an information set allowing the unambiguous and complete description of all causal relations defining the system functioning. It proposes that the dynamics of y_m consists of a portion that can be explained by the information set after excluding y_k plus an unexplained part. If y_k is helpful to forecast to a certain extent the unexplained part, y_k is said to Granger-cause y_m. In our application, the information set is formed by SAP, HP, and R. One of the reasons for this choice is that SAP, HP, and R are relevant cardiovascular parameters carrying information about the whole cardiovascular system and, in particular, about some notable subsystems (i.e., vasculature, heart, and respiratory system). The second reason is that relevant causal interactions between these variables are expected (e.g., SAP is expected to drive HP through the baroreflex). It is worth stressing that the number of signals forming the information set cannot be reduced further in cardiovascular variability studies. Indeed, the exclusion of HP or SAP from the information set would prevent the study of closed-loop

HP-SAP interactions, while the exclusion of R might introduce important bias in causality analysis by leading to the erroneous attribution of a significant fraction of HP variability to the baroreflex (Porta et al., 2012). Since this information set remains quite limited in terms of representation of the causal relations in the cardiovascular system, we can only detect "prima facie" causal relations (Granger, 1980).

12.5.3   GRANGER CAUSALITY APPROACH IN THE TIME DOMAIN

Granger causality approaches have been devised in the time, frequency, and information domains. The frequency domain approaches to Granger causality are probably the ones most frequently utilized (Kamiński and Blinowska, 1991; Sameshima and Baccalá, 1999; Baccalá and Sameshima, 2001; Kamiński et al., 2001; Nedungadi et al., 2001; Porta et al., 2002; Albo et al., 2004). The most important advantage of the frequency domain Granger causality methods is the possibility of testing causality in specific frequency bands without any need to band-pass filter the series (a procedure leading to the destruction of possible nonlinear causal relations). Since short-term cardiovascular regulatory mechanisms operate in the LF and HF bands, frequency domain approaches were exploited in cardiovascular studies (Porta et al., 2002; Nollo et al., 2005). Unfortunately, frequency domain approaches have a notable drawback. The distribution of the frequency domain causality indexes under the null hypothesis of the absence of a causal relation is not a priori known, and therefore, it must be built empirically according to computer-intensive bootstrapping procedures or Monte Carlo methods (e.g., the surrogate data approach). However, this approach is extremely time-consuming compared to the F-test and raises the issue of the best method for constructing surrogates consistent with the null hypothesis of absence of causality (Faes et al., 2010). Causality approaches in the information domain are more recently devised (Faes et al., 2001; Paluš and Stefanovska, 2003; Hlaváčková-Schindler et al., 2007; Vejmelka and Paluš, 2008; Vlachos and Kugiumtzis, 2010). Information domain Granger causality techniques have a relevant advantage compared to time and frequency domain approaches: they do not assume linear interactions among the series, thus being particularly attractive when nonlinear interactions are present. However, even in case of information domain methods, the distribution of the causality indexes under the null hypothesis of absence of causal relation is not a priori given and a surrogate data approach is mandatory to check the significance of the causal relation.

In the present study, a time-domain approach based on predictability improvement was exploited to assess Granger causality (Granger, 1980). Given a series y_m belonging to the information set, the approach is based on the quantification of the levels of unpredictability of y_m given the full information set and given a reduced information set obtained from the original one after excluding the signal supposed to be the cause. If the levels of unpredictability, assessed in the time domain as the variance of the prediction error, are significantly different according to the F-test, a significant causal relation is detected. The use of the F-test allows a straight and easy check of the significance of the causal relation simply by assessing the probability of having

F-values larger than that derived from the series. If this value is marginal (i.e., the probability of type I error is negligible), then a causal relation is robustly detected.

12.5.4 CAUSAL INTERACTIONS BETWEEN HP AND SAP DURING A GRADED HEAD-UP TILT PROTOCOL

One of the most important cardiovascular regulatory mechanisms is based on the bidirectional causal relation between HP and SAP. Indeed, SAP variations, sensed by baroreceptors, induces via neural circuits HP modifications counteracting SAP changes (Smyth et al., 1969). In turn, HP variations cause SAP changes depending on ventricular filling, cardiac contractility, and diastolic runoff. This study confirms the importance of bidirectional HP-SAP interactions at REST and during the entire experimental protocol. Indeed, the bidirectional (i.e., closed loop) relation between HP and SAP was found in a percentage of subjects close to 100 (i.e., 84 at REST) and this percentage remained high and constant during the entire experimental protocol. The identified HP-SAP bidirectional interactions could not be considered the result of the biasing effect of respiratory influences affecting both SAP and HP. Indeed, according to Porta et al. (2012), R was explicitly taken into account in the universe of knowledge utilized to describe HP-SAP causality, thus preventing the biasing effect on the estimation of causality frequently observed when a third signal (here R) is responsible for a large part of the correlation found between a given pair of series (here HP and SAP). It is worth pointing out that, given the large presence of HP-SAP bidirectional interactions, the estimate of the baroreflex gain based on spontaneous HP and SAP variabilities during a head-up tilt requires model-based closed-loop approaches (Porta et al., 2009) instead of cross-spectral techniques (Cooke et al., 1999) assuming an open-loop relation.

When the dominant direction of causality was inferred based on a directionality index, we found that it was mainly from HP to SAP at REST and at low tilt table inclinations, whereas it shifted toward the reverse causal direction (i.e., from SAP to HP) at the highest tilt table angles. This result is similar to that obtained over the same experimental protocol through a conditional entropy approach (Porta et al., 2011b). In contrast to Porta et al. (2011b), in this study the directionality index between SAP and HP was assessed by accounting for *R*, thus confirming that the biasing effect of *R* on the estimation of HP-SAP causality was negligible, probably in relation to the progressive decoupling between HP and *R* induced by sympathetic activation (Montano et al., 1994; Cooke et al., 1999; Furlan et al., 2000) and to the limited magnitude of the perturbing action of R on SAP (Cooke et al., 1999). The observed shift of the dominant causality from HP to SAP at low tilt table angles to the reverse causal direction at the highest tilt table inclination suggests an interesting strategy that might be followed by cardiovascular control to select the most helpful causal relations to face a given situation. The autonomic nervous system might select the dominant causality pattern in closed-loop interactions according to the level of autonomic activity and/or modulation. Indeed, when the sympathetic tone is high (i.e., at high tilt table angles), cardiovascular control selects the baroreflex (i.e., the causal relation from SAP to HP) as the preeminent causal relation in an attempt to limit SAP changes, while when the sympathetic tone is low (i.e., at low tilt table angles), it privileges, as a dominant causal link, the feedforward coupling to contribute to SAP changes and sustain arterial pressure levels. This strategy has the advantage of preserving the closed-loop HP-SAP regulation while providing flexibility in the presence of changeable conditions and quickness of reaction in risky situations.

12.5.5   ASSESSING CAUSAL INTERACTIONS INVOLVING R DURING A GRADED HEAD-UP TILT PROTOCOL

The influence of R on HP is mainly the result of the direct coupling between respiratory centers and autonomic efferent activities, activation of cardiopulmonary reflex, and direct stimulation of the sinus node tissue (Vatner and Zimpfer, 1981; Gilbey et al., 1984; Hakumaki, 1987; Bernardi et al., 1989; Saul et al., 1991; Eckberg, 2003). However, in this study, even the reverse causal link was observed (i.e., from HP–R). Several research papers reported that central respiratory drives, inducing changes of HP via variations of vagal firing, induced delayed effects on R when measured through inductance plethysmographic methods based on thoracic belts, thus suggesting a causal link from HP to R (Saul et al., 1989; Yana et al., 1993; Perrott and Cohen, 1996). This study supports the hypothesis of bidirectional interactions between HP and R, thus emphasizing that the causal link from HP to R cannot be dismissed when modeling the HP–R interactions. The percentage of subjects with bidirectional causality between R and HP was significant at REST (i.e., 63), and this percentage decreased as a function of the tilt table inclination. This finding implies that autonomic activity and/or modulation plays an important role in governing HP–R causality. This result can be explained in terms of the decrease of respiratory sinus arrhythmia produced by the vagal withdrawal proportional to the magnitude of the gravitational stimulus (Montano et al., 1994; Cooke et al., 1999; Porta et al., 2007). This decrease leads to the progressive uncoupling between R and HP (Porta et al., 2011a). In agreement with this observation, the percentage of subjects with uncoupled HP and R series gradually increased with a tilt table angle. We hypothesize that vagal withdrawal is more likely to be responsible for changes of the HP-R causality relation as a function of the tilt table inclination than sympathetic activation. This hypothesis is based on the observation that the gain of the transfer function from R to HP was significantly reduced during vagal blockade induced by atropine, while it was only slightly affected by the β-sympathetic blockade induced by propranolol (Yana et al., 1993). The incomplete uncoupling between HP and R observed at the highest tilt table angle (R and HP are uncoupled only in 37% of the subjects during T90) might be the result of the direct (not mediated by vagal firing) effects of respiratory changes of intrathoracic pressure on the sinus node tissue (Bernardi et al., 1989).

Respiratory-related fluctuations of intrathoracic pressure induce variations of venous return and, in turn, of stroke volume contributing to fluctuations of SAP values at the respiratory rate (Innes et al., 1993; Toska and Eriksen, 1993; Caiani et al., 2000), thus implying a causal relation from R to SAP. Accordingly, in this study, unidirectional interactions from R to SAP were found frequently (i.e., in 68% of the subjects). The percentage of subjects with unidirectional causality from R to SAP was unaffected by the tilt table angle, thus suggesting that the autonomic nervous system plays a negligible role in the modulation of causality from R to SAP. This result is in agreement with previous observations indicating that the magnitude of the relation between R and SAP is not affected by autonomic blockades (Saul et al., 1991). At difference with the HP-R causal interactions, bidirectional interactions were found in a significantly smaller fraction of subjects than unidirectional ones from R to SAP. Therefore, the assumption that R is an exogenous source for SAP (i.e., R affects SAP without being affected) holds better than the same postulation on HP. The causal link from SAP to R was less detected because the fast neural actions were not involved in the SAP-R relation, as suggested by the independence of the percentage of SAP-R causality schemes on the tilt table angle.

12.6   CONCLUSIONS

This study demonstrates that causality analysis can provide information about cardiovascular control that cannot be derived from traditional tools such as spectral and cross-spectral methods, and even from more sophisticated approaches based on multivariate modeling. We propose that a simple Granger causality approach in the time domain based on the concept of predictability improvement might be sufficient to study the closed-loop interactions between HP and SAP and typify the relation with R. The study suggests that cardiovascular control plays an active role in governing causality patterns and uses the strategy of selecting causality patterns to pick up the most appropriate cardiovascular regulatory mechanism. We found that cardiovascular control can modulate the dominance of a causal relation in closed-loop interactions and deactivate causal links via the imposition of uncoupling among variables. Additional studies should be carried out to test different Granger causality approaches based on frequency and information domains to better clarify whether they can add relevant information that cannot be derived from the current approach in the time domain. The application of methods capable of describing nonlinear causal interactions (in this study, the approach is fully linear) might be helpful to clarify the role played by nonlinearities in cardiovascular variability interactions.

REFERENCES

Akaike, H. 1974. A new look at the statistical novel identification. *IEEE Trans. Autom. Control* *19*: 716–723.

Akselrod, S., D. Gordon, F. A. Ubel et al. 1981. Power spectrum analysis of heart rate fluctuations: A quantitative probe of beat-to-beat cardiovascular control. *Science 213*: 220–223.

Albo, Z., G. V. Di Prisco, Y. Chen et al. 2004. Is partial coherence a viable technique for identifying generators of neural oscillations? *Biol. Cybern. 90*: 318–326.

Baccalá, L. A. and K. Sameshima. 2001. Partial directed coherence: A new concept in neural structure determination. *Biol. Cybern. 84*: 463–474.

Badra, L. J., W. H. Cooke, J. B. Hoag et al. 2001. Respiratory modulation of human autonomic rhythms. *Am. J. Physiol. 280*: H2674–H2688.

Baselli, G., S. Cerutti, F. Badilini et al. 1994. Model for the assessment of heart period and arterial pressure variability interactions and respiratory influences. *Med. Biol. Eng. Comput. 32*: 143–152.

Baselli, G., A. Porta, O. Rimoldi et al. 1997. Spectral decomposition in multichannel recordings based on multivariate parametric identification. *IEEE Trans. Biomed. Eng. 44*: 1092–1101.

Bernardi, L., F. Keller, M. Sanders et al. 1989. Respiratory sinus arrhythmia in the denervated human heart. *J. Appl. Physiol. 67*: 1447–1455.

Caiani, E., M. Turiel, S. Muzzupappa et al. 2000. Evaluation of respiratory influences on left ventricular function parameters extracted from echocardiographic acoustic quantification. *Physiol. Meas. 21*: 175–186.

Cohen, M. A. and J. A. Taylor. 2002. Short-term cardiovascular oscillations in man: Measuring and modelling the physiologies. *J. Physiol. 542*: 669–683.

Cooke, W. H., J. B. Hoag, A. A. Crossman et al. 1999. Human responses to upright tilt: A window on central autonomic integration. *J. Physiol. 517*: 617–628.

De Boer, R. W., J. M. Karemaker, and J. Strackee. 1985. Relationships between short-term blood pressure fluctuations and heart rate variability in resting subjects I: A spectral analysis approach. *Med. Biol. Eng. Comput. 23*: 352–358.

De Boer, R. W., J. M. Karemaker, and J. Strackee. 1987. Hemodynamic fluctuations and baroreflex sensitivity in humans: A beat-to-beat model, *Am. J. Physiol. 253*: H680–H689.

Eberts, R. W. and B. M. Steece. 1984. A test for Granger-causality in a multivariate ARMA model. *Empir. Econ. 9*: 51–58.

Eckberg, D. L. 1976. Temporal response patterns of the human sinus node to brief carotid baroreceptor stimuli. *J. Physiol. 258*: 769–782.

Eckberg, D. L. 2003. The human respiratory gate. *J. Physiol. 548*: 339–352.

Eckberg, D. L. and J. M. Karemaker. 2009. Point:Counterpoint: Respiratory sinus arrhythmia is due to a central mechanism vs. respiratory sinus arrhythmia is due to the baroreflex mechanism. *J. Appl. Physiol. 106*: 1740–1744.

Faes, L., G. Nollo, and A. Porta. 2001. Information-based detection of nonlinear Granger causality in multivariate processes via a nonuniform embedding technique. *Phys. Rev. E 83*, 051112.

Faes, L., A. Porta, and G. Nollo. 2010. Testing frequency-domain causality in multivariate series. *IEEE Trans. Biomed. Eng. 57*: 1897–1906.

Furlan, R., A. Porta, F. Costa et al. 2000. Oscillatory patterns in sympathetic neural discharge and cardiovascular variables during orthostatic stimulus. *Circulation 101*: 886–892.

Gilbey, M. P., D. Jordan, D. W. Richter et al. 1984. Synaptic mechanisms involved in the inspiratory modulation of vagal cardio-inhibitory neurones in the cat. *J. Physiol. 356*: 65–78.

Granger, C. W. J. 1969. Investigating causal relations by econometric models and cross-spectral methods. *Econometrica 37*: 424–438.

Granger, C. W. J. 1980. Testing for causality. A personal viewpoint. *J. Econ. Dyn. Control 2*: 329–352.

Guzzetti, S., M. T. La Rovere, G. D. Pinna et al. 2005. Different spectral components of 24h heart rate variability are related to different modes of death in chronic heart failure. *Eur. Heart J. 26*: 357–362.

Hakumaki, M. O. K. 1987. Seventy years of the Bainbridge reflex. *Acta Physiol. Scand. 130*: 177–185.

Hlaváčková-Schindler, K., M. Paluš, M. Vejmelka et al. 2007. Causality detection based on information-theoretic approaches in time series analysis. *Phys. Rep. 441*: 1–46.

Innes, J., S. De Cort, W. Kox et al. 1993. Within-breath modulation of left ventricular function during normal breathing and positive-pressure ventilation in man. *J. Physiol. 460*: 487–502.

Kamiński, M. and K. J. Blinowska. 1991. A new method of the description of the information flow in the brain structure. *Biol. Cybern. 65*: 203–210.

Kamiński, M., M. Ding, W. A. Truccolo et al. 2001. Evaluating causal relations in neural systems: Granger causality, directed transfer function and statistical assessment of significance. *Biol. Cybern. 85*: 145–157.

Kara, T., K. Narkiewicz, and V. K. Somers. 2003. Chemoreflexes—Physiology and clinical implications. *Acta Physiol. Scand. 177*: 377–384.

Kay, S. M. 1989. *Modern Spectral Analysis: Theory and Application.* Englewood Cliffs: Prentice-Hall.

Kim, T.-S. and M. C. K. Khoo. 1997. Estimation of cardiorespiratory transfer under spontaneous breathing conditions: A theoretical study. *Am. J. Physiol. 273*: H1012–H1023.

Kitney, R. I., T. Fulton, A. H. McDonald et al. 1985. Transient interactions between blood pressure, respiration and heart rate in man. *J. Biomed. Eng. 7*: 217–224.

Kleiger, R. E., J. P. Miller, J. T. Bigger et al. 1987. Decreased heart rate variability and its association with increased mortality after acute myocardial infraction. *Am. J. Cardiol. 59*: 256–262.

La Rovere, M. T., G. D. Pinna, S. H. Hohnloser et al. 2001. Baroreflex sensitivity and heart rate variability in the identification of patients at risk for life-threatening arrhythmias. *Circulation 103*: 2072–2077.

La Rovere, M. T., G. D. Pinna, R. Maestri et al. 2003. Short-term heart rate variability strongly predicts sudden cardiac death in chronic heart failure patients. *Circulation 107*: 565–570.

Laude, D., J. L. Elghozi, A. Girard et al. 2004. Comparison of various techniques used to estimate spontaneous baroreflex sensitivity (The EuroBaVar Study). *Am. J. Physiol. 286*: R226–R231.

Lütkepohl, H. 2005. *New Introduction to Multiple Time Series Analysis.* Berlin, Heidelberg: Springer-Verlag.

Malliani, A., M. Pagani, F. Lombardi et al. 1991. Cardiovascular neural regulation explored in the frequency domain. *Circulation 84*: 482–492.

Montano, N., T. Gnecchi-Ruscone, A. Porta et al. 1994. Power spectrum analysis of heart rate variability to assess changes in sympatho-vagal balance during graded orthostatic tilt. *Circulation 90*: 1826–1831.

Mukkamala, R., K. Toska, and R. J. Cohen. 2003. Noninvasive identification of the total peripheral resistance baroreflex. *Am. J. Physiol. 284*: H947–H959.

Mullen, T. J., M. L. Appel, R. Mukkamala et al. 1997. System identification of closed loop cardiovascular control: Effects of posture and autonomic blockade. *Am. J. Physiol. 272*: H448–H461.

Nedungadi, A.G., M. Ding, and G. Rangarajan. 2001. Block coherence: A method for measuring the interdependence between two blocks of neurobiological time series. *Biol. Cybern. 104*: 197–207.

Nollo, G., L. Faes, A. Porta et al. 2002. Evidence of unbalanced regulatory mechanism of heart rate and systolic pressure after acute myocardial infarction. *Am. J. Physiol. 283*: H1200–H1207.

Nollo, G., L. Faes, A. Porta et al. 2005. Exploring directionality in spontaneous heart period and systolic arterial pressure variability interactions in humans: Implications in the evaluation of baroreflex gain. *Am. J. Physiol. 288*: H1777–H1785.

Pagani, M., F. Lombardi, S. Guzzetti et al. 1986. Power spectral analysis of heart rate and arterial pressure variabilities as a marker of sympatho-vagal interaction in man and conscious dog. *Circ. Res. 59*: 178–193.

Paluš, M. and A. Stefanovska. 2003. Direction of coupling from phases of interacting oscillators: An information-theoretic approach. *Phys. Rev. E 67*, 055201.

Patton, D. J., J. K. Triedman, M. H. Perrott et al. 1996. Baroreflex gain: Characterization using autoregressive moving average analysis. *Am. J. Physiol. 270*: H1240–H1249.

Perrott, M. H. and R. J. Cohen. 1996. An efficient approach to ARMA modeling of biological systems with multiple inputs and delays. *IEEE Trans. Biomed. Eng. 43*: 1–14.

Pincus, S. M., T. R. Cummins, and G. G. Haddad. 1993. Heart rate control in normal and aborted-SIDS infants. *Am. J. Physiol. 33*: R638–R646.

Porta, A., F. Aletti, F. Vallais et al. 2009. Multimodal signal processing for the analysis of cardiovascular variability. *Philos. Trans. R. Soc. A 367*: 391–408.

Porta, A., G. Baselli, O. Rimoldi et al. 2000a. Assessing baroreflex gain from spontaneous variability in conscious dogs: Role of causality, and respiration. *Am. J. Physiol. 279*: H2558–H2567.

Porta, A., T. Bassani, V. Bari et al. 2012. Accounting for respiration is necessary to reliably infer Granger causality from cardiovascular variability series. *IEEE Trans. Biomed. Eng. 59*: 832–841.

Porta, A., T. Bassani, V. Bari et al. 2011a. Model-based assessment of baroreflex and cardiopulmonary couplings during graded head-up tilt. *Comput. Biol. Med. 42*: 298–305.

Porta, A., A. M. Catai, A. C. M. Takahashi et al. 2011b. Causal relationships between heart period and systolic arterial pressure during graded head-up tilt. *Am. J. Physiol. 300*: R378–R386.

Porta, A., R. Furlan, O. Rimoldi et al. 2002. Quantifying the strength of the linear causal coupling in closed loop interacting cardiovascular variability signals. *Biol. Cybern. 86*: 241–251.

Porta, A., S. Guzzetti, N. Montano et al. 2000b. Information domain analysis of cardiovascular variability signals: Evaluation of regularity, synchronisation and co-ordination. *Med. Biol. Eng. Comput. 38*: 180–188.

Porta, A., E. Tobaldini, S. Guzzetti et al. 2007. Assessment of cardiac autonomic modulation during graded head-up tilt by symbolic analysis of heart rate variability. *Am. J. Physiol. 293*: H702–H708.

Richman, J. S. and J. R. Moorman. 2000. Physiological time-series analysis using approximate entropy and sample entropy. *Am. J. Physiol. 278*: H2039–H2049.

Riedl, M., A. Suhrbier, H. Stepan et al. 2010. Short-term couplings of the cardiovascualr system in pregnant women suffering from pre-eclampsia. *Philos. Trans. R. Soc. A 368*: 2237–2250.

Rosenblum, M. G. and A. Pikovsky. 2001. Detecting direction of coupling in interacting oscillators. *Phys. Rev. E 63*, 045202.

Sameshima, K. and L. A. Baccalá. 1999. Using partial directed coherence to describe neuronal ensemble interactions. *J. Neurosci. Methods 94*: 93–103.

Saul, J. P., R. D. Berger, P. Albrecht et al. 1991. Transfer function analysis of the circulation: Unique insights into cardiovascular regulation. *Am. J. Physiol. 261*: H1231–H1245.

Saul, J. P., R. D. Berger, M. H. Chen et al. 1989. Transfer function analysis of autonomic regulation II. Respiratory sinus arrhythmia. *Am. J. Physiol. 256*: H153–H161.

Smyth, H. S., P. Sleight, and G. W. Pickering. 1969. Reflex regulation of the arterial pressure during sleep in man. A quantitative method of assessing baroreflex sensitivity. *Circ. Res. 24*: 109–121.

Söderström, T. and P. Stoica. 1988. *System Identification.* Englewood Cliffs: Prentice-Hall.

Task Force of the European Society of Cardiology and the North American Society of Pacing and Electrophysiology. Standard of measurement, physiological interpretation and clinical use. 1996. *Circulation 93*: 1043–1065.

Task Force on Sudden Cardiac Death on the European Society of Cardiology 2001. *Eur. Heart J. 22*: 1374–1450.

Toska, K. and M. Eriksen. 1993. Respiration-synchronous fluctuations in stroke volume, heart rate and arterial pressure in humans. *J. Physiol. 472*: 501–512.

Vatner, S. F. and M. Zimpfer. 1981. Bainbridge reflex in conscious, unrestrained, and tranquilized baboons. *Am. J. Physiol. 240*: H164–H167.

Vejmelka, M. and M. Paluš. 2008. Inferring the directionality of coupling with conditional mutual information. *Phys. Rev. E 77*, 026214.

Vlachos, I. and D. Kugiumtzis. 2010. Nonuniform state-space reconstruction and coupling direction. *Phys. Rev. E 82*, 016207.

Xiao, X., T. J. Mullen, and R. Mukkamala. 2005. System identification: A multi-signal approach for probing neural cardiovascular regulation. *Physiol. Meas. 26*: R41–R71.

Yana, K., J. P. Saul, R. D. Berger et al. 1993. A time domain approach for the fluctuation analysis of heart rate related to instantaneous lung volume. *IEEE Trans. Biomed. Eng. 40*: 74–81.

Section IV

Epilogue

13 Multivariate Time-Series Brain Connectivity
A Sum-Up

Luiz A. Baccalá and Koichi Sameshima

CONTENTS

13.1 DIRECTEDNESS VICTORIOUS

To provide a panorama of the field as fair and wide as possible, we as editors have striven to impose as few changes of content as possible to the chapters. One may appreciate this freedom by comparing what authors have to say about the relationship between Granger causality and the frequency domain quantities of Chapters 2 and 4. Rather than using or imposing our opinion in this and other matters, we believe it is important to have views presented freely and defer decisions to the reader who is the ultimate judge.

Because the emphasis here has been on the general multivariate time-series case, a basic truth may have been insufficiently stressed—the fact that in the bivariate case ($N = 2$), all concepts coalesce, and DTF and PDC are strictly equal and reflect the exact frequency domain decomposition of bivariate G-causality.

Even in the case of just two time-series ($N = 2$), we hope the reader has been able to appreciate the benefits of exposing the direction of interactions, something that usual correlation/coherence-based approaches fail to do. Use of G-causality (and DTF/PDC in contexts where frequency domain interpretations are meaningful) is already a huge step beyond ordinary correlation/coherence methods.

In fact, from now on, undirected connectivity descriptions, even in the bivariate case, cannot but be considered poor and inadequate and should be replaced by directed ones, even if $N = 2$. To appreciate this subtle point, the reader may examine what happens in Figure 7.6 where the peak spectral activity in $x_2(n)$ is subject to high influence from $x_1(n)$ as imposed by model construction, but which nonetheless exhibits fairly low classical coherence at that frequency.

As $N > 2$, more options become available for structural description. It then becomes important that one should take explicit advantage of this brand new realm of possibility.

The issues associated with the notion of instantaneous causality, that is, also employing the present of the other time series for prediction improvement, is a more difficult one and requires either side information (much as in the case of SEM) or exploitation of possibly inherent non-Gaussian signal behavior (Chapter 6) for conclusive estimation, but to some extent it also allows some "directedness" of interaction to be taken into account.

We henceforth propose that from now on connectivity must be considered in its "directed" capacity since, all in all, it is a data-driven property that is intrinsically describable and one large step ahead of mere correlation/coherence considerations.

13.2 PRACTICAL CHALLENGES

Perhaps the biggest challenge for the practical application of the techniques in this book is model adequacy. Linear models are traditional (Chapter 3); most of all, their strengths and caveats are well known. Even areas of traditional time-series model use have not entirely solved many practical issues, which remain the "art" part of the field. Here, we tried to stick mostly to the new directedness "science" part of the field and provide the reader with software means with which to investigate estimation issues.

To give the reader some notion of the problems we have faced in dealing with data over the years, we mention:

1. Inadequate signal preprocessing
2. Model order issues, and perhaps rather counterintuitively
3. The inappropriate choice of sampling frequencies that are much higher than necessary, in addition to
4. The presence of outlier contaminants
5. Acquisition channel calibration problems.

Even though filtering all data channels with the same filter does not theoretically impact DTF or PDC, filtering transients over short data segments often leads to misestimated connectivity.

In several chapters, the reader will have noticed mention of model order choice criteria with Akaike's being by far the most popular. Obedience to some acceptable measure of "whiteness" of signal residuals (estimated prediction errors) is as important to ensure capture of the dynamics (see Chapter 3). In some cases, when linear models prove inadequate, one must go on to check more closely for the possibility of improved prediction via nonlinear models (Chapter 8) or the need to examine the data from a time-variant model perspective (Chapter 9).

It is interesting to observe that many practical published uses of the techniques presented here have mostly left model quality checks implicit. Much of their trustworthiness could be improved if quality checks were systematically included in publications. Examples of this kind of analysis from beginning to end are presented in Chapters 3 and 4.

Perhaps even more counterintuitive is the fact that very high sampling rates lead to models with many parameters. Parameter profusion introduces estimation inaccuracies that often impact connectivity estimation. In our experience, good parametric spectral estimates are best when signal power spreads itself over the frequency domain between 0 and one half of the sampling rate f_s. If the experimental sampling rate is too high with respect to the maximum signal frequency, the spectrum becomes too narrowly confined and leads to models of high order and low accuracy.

Another context of care with regard to sampling frequency choice is the proper account of signal delays. This is important since causality is only adequately captured if the chosen model order p times the sampling interval $T_s(= 1/f_s)$ is larger than the largest delay between channels so that p will necessarily be high if f_s is high. This leads once more to many parameters and overall estimation performance reduction if the number of observed data points over time is not large enough.

If the signals of interest are stationary only over small periods, the number of required descriptive parameters can be critical since most model estimation algorithms either presume stationarity or only mild departures from it. In practice, there is a trade-off between the number of necessary (statistically stationary) data points and the number of parameters that make up the model. Without too much justification, Marple Jr. (1987) suggests that the total number of observed samples should be at least three times the number of parameters for a minimally adequate fit.

Lastly, also in practice, one of the most overlooked problems lies with instrumentation: if measurements are to reflect the actual structure of interest, the acquisition channels must have identical delays. Failure to calibrate channel delays can easily produce results that reflect more of the imperfections of the measurement devices than the actual biology of interest since the present inference notions rest heavily on accounting for signal precedence.

It is only when a good linear time-series model can be fit to the available data that one can safely apply the results in this book as described from Chapters 2 through 5 and especially Chapter 7, which treats the associated statistical problems. By a good model, we mean one whose residuals are close to white with as few parameters as possible. It is exactly in the spirit of data reduction that one can interpret the results of Chapter 10 where the measurement of many time series in the so-called (scalp) electrode space is replaced by hopefully more physiologically meaningful representatives of the phenomenon that moreover requires fewer parameters for description. If the relevant dynamics consists of a model order p to describe the total signal delay going from the electrode space, its pN_e^2 free model parameters are reduced to pN_d^2 quantities where the number of electrodes N_e is usually very much larger than the number of representative dipoles N_d.

13.3   A NEW PARADIGM FOR CONNECTIVITY DISCUSSIONS

Last but not least, there is one central and very delicate point we would like to discuss surrounding the current connectivity literature status. Much of it as it now stands distinguishes two main sorts of connectivity (see Friston, 1994, for a review):

1. *Functional connectivity*: "temporal correlation between spatially remote neurophysiological events."
2. *Effective connectivity*: "the influence one neuronal system exerts over another."

This classification is fine if only undirected notions of connectivity through correlation/coherence are available. Furthermore, one must also not forget that connectivity associated to the observed dynamics takes place within the background of structural, that is, anatomical, connections which make up the substrate to which the physiologically *operating* mechanisms must ultimately conform.

The last paragraph's keyword is "operating" since in connectivity analysis the goal is to ultimately appraise when and whether an existing structure (neural projection) is *active*. Obviously, an active connection implies "effective" connectivity and one would expect there to be evidence of "functional" connectivity as well. On first thoughts, it is on this that the current paradigm rests. These considerations have a weak point though, for it is possible to envisage an all too simple counterexample.

Consider the three structures investigated in the illustrations of Chapter 5, when, instead of Equation 5.36, one employs the equation

$$x_3(n) = \gamma x_1(n-2) + \beta x_2(n-1) + w_3(n) \qquad (13.1)$$

leading to the connectivity summarized in Figure 13.1, where $x_1(n)$ now has direct connections to both $x_2(n)$ and $x_3(n)$ and $x_2(n)$'s signal also flows to $x_3(n)$ directly.

If $\gamma = -\alpha\beta$, it is easy to show that the coherence between $x_1(n)$ and $x_3(n)$ is identically zero regardless of the frequency, that is, $Coh_{31}(f) = 0$ because $DTF_{31}(f) = 0$ as a result of the additional effect of the direct link between $x_1(n)$ and $x_3(n)$ has $(PDC_{31}(f) \neq 0)$ by counterbalancing the indirect phase inverting signal through x_2. This translates into zero net influence from $x_1(n)$ onto $x_3(n)$ and hence no effective total connection. Coherence nullity, by definition, implies that there is no "functional" connectivity between $x_1(n)$ and $x_3(n)$ either. Nevertheless, the structures are physically connected by construction, and this condition is nevertheless fully detectable. This example shows that the functional/effective connectivity vocabulary is insufficient to describe the actual underlying situation. All connections in this example are active; tampering with anyone of them would alter the overall dynamics and have clear experimentally detectable consequences. What this example attempts to show

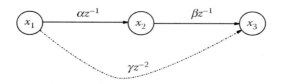

FIGURE 13.1 The addition represented by the new term in Equation 13.1 of the direct link (dashed arrow) from x_1 to x_3 if properly weighted $\gamma = -\alpha\beta$ balances out the indirect signal effect that propagates from x_1 through x_2. The integer exponents of the z variables in the diagram represent the delays in terms of the number of samples.

TABLE 13.1
Key to the Relationship between Connectivity
Quantifiers and Its Classification

	Direct	Indirect
Active	$PDC \neq 0$	$PDC = 0$ and $DTF \neq 0$
Inactive	$PDC = 0$	$DTF = 0$

is that the usual definitions of functional and effective connectivity, as they stand, are too vague, unsatisfactory, and simply neither apply nor convey the full connectivity picture.

We therefore propose a change in how to classify connections. Because quantifiers such as PDC allow detection of direct connections and DTF allows detection of links directly or otherwise, we propose to distinguish between *active* and *inactive* connections, with the additional qualifier of stating whether connectivity is:

1. *Direct*: because it takes place without intervening structures, or
2. *Indirect*: because it takes place through intermediary structures.

A key to this classification could follow the inference associated with PDC and DTF results as shown in Table 13.1.

The frequency selective nature of connections as revealed in Figure 7.6 further supports the need to stress the properties of the connections themselves, how they are active, and under what conditions.

The current mainstream status of connectivity applications is built upon the ideas of functional/effective notions, with little, if anything, of directedness descriptions having been adequately explored (Sporns, 2011). Most of the current graph theoretical descriptions implicit in current applications do not explicitly incorporate directedness, whose inclusion seems to open up a number of extra possibilities. Among them, it may lead to more natural ways of parceling out structures associated with resting state networks (RSN) through specific directed network descriptors (Vieira et al., 2012).

13.4   FUTURE BUT URGENT DEVELOPMENTS

Our newly proposed classification of connectivity into active/inactive–direct/indirect connections in Table 13.1 is made possible because the (linear) connectivity detection problem is essentially solved (a manuscript paralleling the developments of Chapter 7 for DTF is in preparation).

The connectivity magnitude problem, on the other hand, is another story. There are many open issues. Even in the case of linear interactions, there are many ways to measure it. One is free to use members of the DTF family (DC, ιDTF) or the PDC family (gPDC, ιPDC, Schelter's et al. rPDC) or hybrid measures like Geweke's or dDTF. Naturally, DTF and PDC differ in the general case* and provide complementary descriptions and discussion as to which is more helpful in addressing a given physiologic question, which is in order.

Provision of quantities such as ιPDC possessing rigorous confidence intervals and accounting for size effects opens the way for systematizing experimental design involving intra subject, extra subject, intra group, and extra group comparisons.

Because the methods herein rely so heavily on the precise acquisition of simultaneous data, a thorough review of protocols is needed in some data-acquisition modalities. This is especially true of multichannel EEG equipment with its many existing brands whose design calls for synchronized channel acquisition that is sometimes unavailable in cases of cheaper systems. Standard equipment calibration protocols are required to ensure reliable use of the current approaches.

The purpose of the AsympPDC package included in this book is chiefly didactic and maps Chapter 7 equations closely; work is under way to optimize the programs and offer versions that may be incorporated into more general analysis packages.

Naturally, further research into connectivity to describe nonlinear interactions is important (Chapter 8) with issues of reliability, record length, where the nonlinear detection problem remains in practice still only lightly explored. Extending the studies of nonlinear effects together with the incorporation of time-variant estimation concepts (Chapter 9) are natural pathways into future investigations.

We would like to reinforce the need to develop reliable means of data/model reduction of which the scalp-to-electrode space approach adopted in Chapter 10 is but one example. This is especially important because limitations of signal duration versus the number of available time-series will always exist and need to be addressed.

Given PDC's (or equivalently G-Causality's as well) ability to allow directed graph connectivity structure representations, the issue of systematic directed graph analysis comes to the fore (see how difficult it is to make much sense of the patterns in Figure 7.13). We made some early attempts through an early analysis of multipatient ictal observations (Baccalá et al., 2004), but it is clear that the resulting graphs require specific summarization tools for analysis and that plotting large connectivity matrices is hardly the way to go in facilitating interpretation. A recent review is provided by de Vico Fallani and Babiloni (2010) without employment of the new connectivity classification criteria from Section 13.3.

An important application goal of this kind of analysis is their employment as translational medicine tools (Cosmatos and Chow, 2008). While it is important to provide reliable inference tools, one must not forget that cognitively meaningful visualization and summarization tools for presenting the results are also vital and must be developed if the ideas herein are one day to find their way into useful clinical practice.

13.5   FINAL REMARKS

Finally, since you, the reader, are the ultimate judge, in that capacity you are invited to use the software provided with this volume, generate artificial data, use data of your

---

* Coincidence of DTF and PDC connectivity graphs implies that all existing active connections are direct.

interest, contaminate it, and play with sample sizes to perceive both the possibilities and the practical difficulties. We hope this will be a rich learning process.

REFERENCES

Baccalá, L. A., M. Alvarenga, K. Sameshima et al. 2004. Graph theoretical characterization and tracking of the effective neural connectivity during episodes of mesial temporal epileptic seizure. *J. Integ. Neurosci.* 3: 379–395.

Cosmatos, D. and S.-C. Chow. 2008. *Translational Medicine: Strategies and Statistical Methods*. Boca Raton, FL: Chapman & Hall/CRC.

de Vico Fallani, F. and F. Babiloni. 2010. *The Graph Theoretical Approach in Brain Functional Networks: Theory and Applications*. San Rafael, CA: Morgan & Claypool Publishers.

Friston, K. 1994. Functional and effective connectivity in neuroimaging: A synthesis. *Hum. Brain Mapp.* 2: 256–278.

Marple Jr, S. 1987. *Digital Spectral Analysis*. Englewood Cliffs, NJ: Prentice-Hall.

Sporns, O. 2011. *Networks of the Brain*. Cambridge: MIT Press.

Vieira, G., J. R. Sato, E. Amaro Jr et al. 2012. Finding fMRI resting-state network (RSN) structures with help of graph hubs and authorities. In *Organization for Brain Mapping OHBM2012*, Beijing.

Index